Raw Earth Solutions for a Common House in Niger

Alisia Tognon | Luca Trabattoni

Necessary Architecture

Necessary Architecture

Raw Earth Solutions for a Common House in Niger

Alisia Tognon
Luca Trabattoni

CRC Press/Balkema is an imprint of the Taylor & Francis Group, an informa business

Typeset by Deanta Global Publishing Services, Chennai, India

Library of Congress Cataloging-in-Publication Data
A catalog record has been requested for this book

Published by: CRC Press/Balkema
Schipholweg 107C, 2316 XC Leiden, The Netherlands
e-mail: Pub.NL@taylorandfrancis.com
www.crcpress.com – www.taylorandfrancis.com

ISBN: 978-0-367-74403-8 (Hbk)
ISBN: 978-0-367-74404-5 (Pbk)
ISBN: 978-1-003-15763-2 (eBook)

DOI: 10.1201/9781003157632

Falwal, Niger ©AT 2016

Table of contents

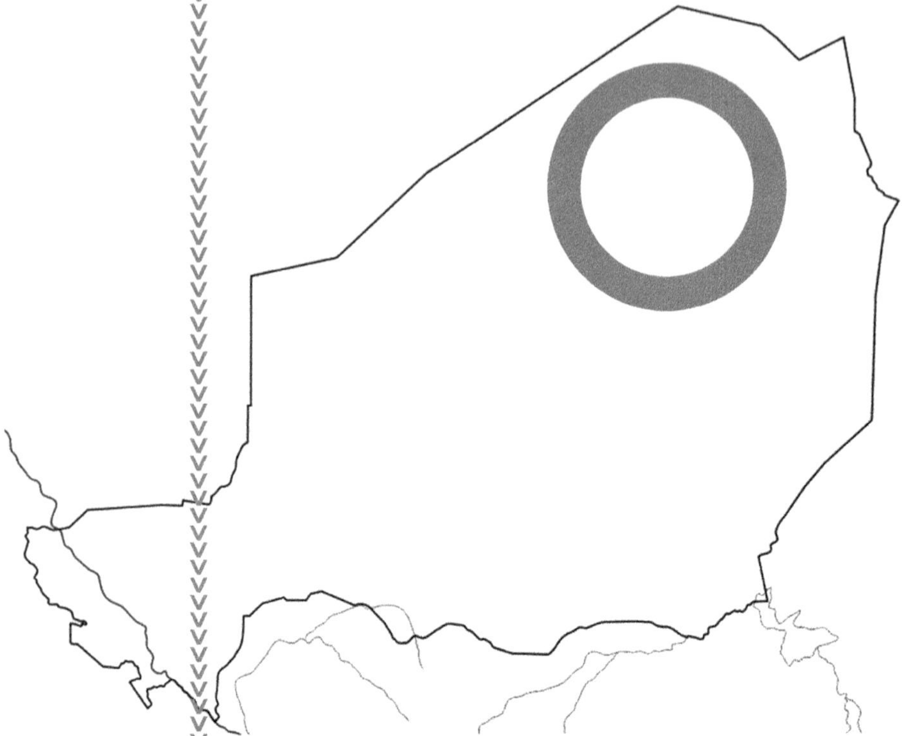

The necessary Niger

Alisia Tognon | Luca Trabattoni

00

DOI: 10.1201/9781003157632-101

Ex africa semper aliquid novi
Plinio il Vecchio

Niger is sand, light and heat. It is the scorching air, which guides the immense deserted land dominating the whole landscape. It is the beaten earth of the roads merged with the earth of the built walls. The black nights, illuminated only by the lights of the stars, are hot and sultry as to seem to dry up sleeping bodies. Shade and water are vital resources for man and animals.

The sand fills the lungs and permeates all; the sun burns every stone; earth colours everything red; sudden sandstorms crush the overwhelming sky; the trees are dry and leafless, although all around plastic bags carried by the wind take the place of the vanished vegetation.

The capital Niamey is shaped into the chaos of disconnected urban growth without planning. Here the intelligibility of volumes, streets and walls blends with the surrounding landscape defined by an endless horizon.

Thinking about Niger as a place for designing means carefully focusing on its fragility and grasping its potential. Niger is a country of which little is said except for the issues related to its poverty (it was the poorest country in the UN world in 2018), the richness of its subsoil exploited by foreign companies, the political instability, due to the weight of foreign nations in its domestic politics, the acts of terrorism and political crises of its neighbours, making it the centre of a crossroads of migratory routes and illegal smuggling of migrants.

However, the fragility of the environment is combined on one side with a rich culture, which triggered, since ancient times, an architectural response to the challenging climatic conditions of the Sahelian context; secondly, the poverty of economic means is linked to the worsening of weather conditions (which causes a constant increase in desertification and deforestation) and the scarcity of resources.

Architecture finds its state of necessity in dealing with similar contexts. The circumstances led us to take an interest in, and commitment to engaging with, increasingly fragile contexts. The emergency, meaning paucity of resources (economic, material and scientific), is the common ground when you are facing a project in these types of areas.[1]

In these contexts the architecture is, on the one hand, an ethical sense, tied to immediate answers to clear questions, and, on the other, is confronted with a need for creative solutions given the lack of resources. Indeed, the lack of traditional building material is one of the most common characteristics of this kind of experience, which led us to explore alternative ways.

The alternative is the combination of "unusual" and traditional factors. The vernacular experience in these situations is a crucial asset for correct solutions to problems, based on the climate and actual resources (Rudofsky 1964: 14). Unfortunately, we often come across human contexts where much of this knowledge has been lost. In Niger, for example, building techniques related to raw earth have created examples of exceptional architecture, very rarely found today. The ability to control measurements, mixtures and construction technology is a fundamental resource to be recovered and preserved.

Indeed, studying history gives a wide range of appropriate solutions, finding then the appropriate technological, cultural and scientific context (Schumacher 1973). The experience of Hassan Fathy (1969), who rediscovers raw earth technologies through experienced craftsmen and applies them in a new context, is, however, an enviabley rare case. More often, in our personal experience, it happens that we propose local techniques or their reinterpretation to local people who ignore them for the most part or entirely.

The issue of lack of material wealth understood as very few resources and limited knowledge, linked to emergency issues, such as migration or famine, is a crucial and critical point for the definition of consistent and positive design solutions in such contexts.

However, contemporary architecture brings with it new spatial requirements, related to the different actual situations (conflicts, hygiene, lifestyle), and therefore requires a comparison with the vernacular technologies that exceed the limits of technique and somehow becomes a linguistic opportunity.

Acquiring awareness is necessary through studying history for the interpretation and renovation of the forgotten tradition. It is essential to integrate the knowledge acquired from the tradition with the contemporary and the theme of recycling and use of waste materials, which become a "challenge" in the emerging countries, where any object and material finds new life and new use until its tangible reduction to a "zero" level.

Therefore the "unusual" element joins the vernacular tradition to find innovative construction solutions, which are based on the principles of low technology and the Blue Economy, which exploit approaches to temperature control and passive ventilation applied to simple and effective technologies.

The necessity of architecture consequently lies in the ability to integrate all of these factors, making a synthesis of spatial construction requirements and giving appropriate responses to the context, economically, technically and ecologically efficiently. The resulting language has a strong identity, in which the material becomes a primary material of construction (Espuelas 2009).

Among the many possible solutions, raw earth is the most versatile and widespread in the world. Its applications are manifold, from bricks to 3D printing. Moreover, its availability is mostly comprehensive, and the efficiency of its performance goes well with contexts like Niger. Alongside the construction methods, however, the cultural aspect linked to the transmission of knowledge is crucial. Knowledge rediscovery becomes a primary form of emancipation in a fragile environment. Thus, the architectural work acquires an additional value when combined with the self-construction processes. In this way, allowing non-experts to build a dwelling, it is possible to give them a new cleverness.

For this reason, professional experience has become even teaching activities in academia and concrete applications of techniques with the students.[2] In addition, professional offices and educational activities have also become a research field on these issues. The emergency conditions, vernacular techniques and recycled materials in local settings have been placed as a key point of studies, both theoretical and practical.

The professional, teaching and research experiences have led to the investigation of these issues, which are a perfect example of application in the specific case of Niger. On the one hand, the chance of a job after the conclusion of a doctoral thesis and other professional participations in similar fragile contexts led to this research in constant balance between the "unusual" factors, anchored in traditions.

Fate had allowed us to establish an ongoing study in a country where the lack of resources is second only to the lack of knowledge related to its potential for building and the improvement of housing conditions.

This circumstance happens thanks to the support of Mission Catholique of Dosso, which for years carried out humanitarian activities in this region, and the productive relations with Don Davide Scalmanini, a missionary from Lodi (Italy), who has lived and worked in Niger for years.

Therefore, this text, which is aware of being a piece in the architectural landscape of potential reactions in the Niger's future, derives from the belief that the vernacular experience in these situations represents a key asset for answers to a correct approach. Besides, according to the climate and these resources, the opportunity represented by the update of vernacular architecture can be a useful example even in non-emergency contexts. The high necessity of these topics supports the teaching inherent in the possibility of confronting a still necessary architecture.

>>> Notes

1 As members of the ARCò architecture and cooperation studio, we have been dealing with design in emergency contexts for years. The issues of sustainability, reuse, recycling, natural materials, the rediscovery of the vernacular and self-construction are at the centre of our professional experience in the field. www.ar-co.org

2 The ARCò School is a nomadic body whose goal is to spread knowledge related to self-construction techniques and recycled materials. The self-building approach switches to multiple situations. Although ARCò school does not have a permanent location, it moves where it is needed, to deal with contingent demands and provide practical solutions or theoretical reflections [http://www.ar-co.org/it/scuola/home/intro/index.php].

Fig. 00.01
Women during a festival.
Dosso, Niger
©AT 2016

The Niger territory
Land between an ethnic crossroads and the dynamism of the desert

01

Alisia Tognon | Lara Spadari

DOI: 10.1201/9781003157632-01

Architecture represents the basic answer to human life on Earth: light, orientation, space, proportions, shape, functions, ornaments, materials, details and construction techniques, combined with a specific type of social organization and a symbolic system, reveal the constant and inevitable relationship between humans and territory.

The evolution of society and urban settlements over the centuries, blended with construction practices, architectural features and technological development, has changed the concept of space and housing practices over time. The historical progress has brought the humans in developed contemporary societies to live in dense urban agglomerations, where the direct link with the territory and nature has disappeared. The difference and the distance between urban areas and natural areas have become increasingly marked, especially in some areas of the world.

Speaking of West Africa, there is a deeply rooted vernacular social organization, which bases its structure on ethnic and spiritual ties. In the past, these societies were based on a segmented system of power or city-states. Therefore, there was no centralized power, but the different ethnic groups took root and created their hegemony in a linear relationship with the territory, cultural traditions and social ties. Furthermore, in these societies it is not possible to isolate political power and economic factors from religious practices and spiritual activities. Social organization is conceived as a great relationship of blood relatives (families), who define various roles and organizations within the villages. The earthly world is reflected in the world of the afterlife, where the human experience converges, like a vast web of relationships. This conception changed when foreigners imposed from outside a partition of the African territory, where the country land is not the ethnic land.

The land on which humans live is part of the social being: it is the original place to be shaped and considered with respect. That meaning is still significant in the Nigerien culture, where it is the Earth that makes man, and everything that man does is linked to it. The territory conceived in the African context acquires a sacred vernacular value because it is the source of fertility and the creator of man (who is generated from clay in African myths) (Lynch, Roberts 2004: 52). It provides humans with the tools to build a space to live. In this sense, vernacular architecture is the combination of divinity and nature, an architecture formulated as a living being, capable of procreating, which changes and possesses a time to exist, which can get sick and die, returning to its primary source and restarting its cycle, regenerating the territory (Bourdier, Minh-ha 2005).

Niger, the myth, the river, the people

Niger is a country born on the European maps following the choices imposed by western colonization. In reality, its origin is more profound and includes a territory with specific geographical/environmental characteristics and different ethnicities. Much more profoundly, it has its origins rooted in a history of mythologies, which find the mother from which everything flows in the environmental characteristics of the territory. Therefore, to understand this land, then it is necessary to recognize its peculiarities and multi-ethnicity which, with specific cultural contributions, defines the architectural characteristics of living.

A legend tells the mythical origin of the peoples of Niger, the legend of Dikko Harakoy (Hockings 1995: 233; Stoller 2009: 133), queen of the waters (Harakoy). Dikko was a mythical figure who belonged to the "Holley" (genes) and protected the inhabitants of the Niger valley. She was a mermaid with the peculiarities of a Peul woman and light-hair, and she used to live in Katakambe, inside the waters of the Niger River. Every evening she went out of the river waters to wait for a new lover and to procreate a line of children from whom different ethnic groups in this region descend today: Kirey, a genius of the wise (Songhai); Mahama Surgu, a nomadic shepherd (Tuareg); Mussa Niawri, hunter (Gurmancé); Hawsakoy Manda, blacksmith (Haussa); Faran Baru Koda, an androgynous genius (Tuareg); Dongo, Dikki's adopted son (Bariba). (Arecchi, Delisse 1999: 35–37; Gallais 1984).

In light of this legend, the relevance of the river is understandable. The Niger River is considered a sacred place because according to the legend of *Dikko Harakoy*, the history of the Niger people originated in its valley.

The nation itself gets its name from the Niger River. The country is a vast land mostly desert, except for the valley, where the river flows. The riverbed occupies a narrow area

Fig. 01.01
Niger River,
Gaya region, Niger
©AT 2016

Fig. 01.02
Family living in Niger River, Gaya Village, Niger
©LS 2017

of the country, on the borders with Burkina Faso, Benin and Nigeria. Due to the presence of water, the agricultural areas and rural villages are concentrated along its riverbanks, and the rural communities derive their livelihood from it.

The Niger River is the third-longest river in Africa, exceeded only by the Nile and Congo Rivers. Moreover, it is the principal river of western Africa, extending over 2,500 miles (about 4,180 km), and its basin has an area of 2,100,000 km^2 (Amani et al. 2015: 39). It runs through Guinea (Loma Mountains), Mali, Niger, on the border with Benin and then through Nigeria, discharging through a massive delta, known as the "Oil River", into the Gulf of Guinea. An uncontaminated oasis extended along the Niger Delta in Nigeria, until colonization by western oil companies (1956). Since then, mangrove forests have given way to oil industries, which are destroying the ecosystem.[1] Today the Niger Delta region (70,000 km^2 with 27 million inhabitants) produces most of the Nigerian oil (Okoh 2006), and the production in 2018 amounted to some 2.05 million barrels per day.[2]

Niger River has always assumed considerable environmental importance on the African continent. It is particularly important to the Niger State. The river is used as a crossroads for transporting goods or people by pirogues. Nevertheless, it also facilitates agricultural activity and vegetation growth by decreasing the drought, which affects most of the country.

Niger is a Sahelian country with an area of 1,267,000 km^2, making it one of the largest countries in West Africa. It is located between 11° 37′ and 23° 33′ latitude North and 0° 06′ and 16° 00′ longitude East, on the South of the Tropic of Cancer. The majority of its territory is desertic, defining it as one of the hottest areas on the planet, marked by a strong climatic instability and by an economical and nutritional high vulnerability.

Even the political and economic situation on the border presents a series of instability (UN 2019). Niger shares a border with Chad in the East, Libya and Algeria in the North, Mali and Burkina Faso in the West and Nigeria and Benin in the South. In recent years the country has been well known as a crossroads for migrants,[3] who from all over West African countries crossed the deserts to reach the Algerian border and from there to cross the Mediterranean to Europe (UNHCR 2019).

In addition to its history of droughts and food insecurity, the analysts marked Niger with three factors, strongly correlated with instability and conflict. Firstly, it is very low-income and landlocked, and, consequently, its governments rely on sizable revenues from mineral exports (Collier 2007). Moreover, rapid population growth produces further stresses. All of these factors add to the complexity of challenges they face in building resilience and ensuring security for their citizens.

The living conditions of most Nigeriens are demanding: the Niger's Human Development Index (HDI) value for 2018 was 0.377,[4] which ranked Niger as last in the 2018 United Nations (UN) Human Development Index (186 out of 186 countries).[5]

To this situation must be added a weak political institution and a fragile national scene, the instability in neighbouring countries and the feeling of dissatisfaction of the Niger citizens.

Like many African states, Niger has experienced a political crisis that perpetuates itself as the consequence of European colonization, specifically French, which still today has created a hidden post economical colonization (Wiafe Amoako 2019). The development of the country has been hampered by military and civil crises. After independence in 1960, the civilian government of Niger was overthrown by the military in April 1974, and the military government lasted until 1990 (Walraven 2003). Between 1990 and 1991, various groups began to get agitated about the civil government and a national conference in 1991 led to establishing the first multi-party constitution and government. Although the governments were generally ineffective, and political and military crises were ongoing (Moestrup 1999). In 1999 a military-controlled transition regime conceded power to a civilian government following free and open presidential and legislative elections in December 1999. However, even today, the political situation is far from stable, also due to religious unrest and to new waves of terrorism in the Sahel region.

Besides, as reported by The World Bank, Niger is heavily dependent on foreign subsidies and loans, which were, on average, roughly equal to national revenues. However, due to internal political instability, external financing has been unstable, going from a low of 4.5 per cent of its gross domestic product (GDP) in 1999 to a high of 17.9 per cent in 2000 (The World Bank 2005: 1).

The population is estimated to be 23,311,000 inhabitants[6] (UN 2019), an annual growth rate of 3.84 per cent (INS 2019). This demographic growth leads to new needs in terms of basic essential services and feeding. Even if the Niger soil is endowed with priceless natural resources which have not yet been exploited, three-quarters of this population lives on the southern fringe more suitable for agriculture, and where rainfall is highest. Only 12 per cent of its area, 46,397 m^2 (FAO 2016), located in its southern part, is favourable to subsistence agriculture which, however, occupies 85 per cent of its population (USAID 2014). Niger is mainly a rural and agrarian economy, but moving to the

North, livelihoods increasingly shift from agriculture to agro-pastoral and pastoralism (USAID 2014). Rain-fed agriculture and livestock herding, done primarily through manual labour, generates about 39 per cent of the country's GDP (FAOSTAT 2011). Primary rain-fed crop production includes millet, sorghum, cotton and corn, while livestock (sheep, goats, camels and cattle) are exported to neighbouring Nigeria. Despite Niger's agrarian character and the critical role of livestock exports, the most considerable portion of the country's export revenues comes from uranium mining based in the north of Agadez (USAID 2014).

There has been a growing interest in investigating the role that ethnic diversity has played in the socio-economic effects. Some studies (Easterly, Levine 1997; Alesina et al. 1997; Mauro 1995) argue that the ethnic diversity explains a substantial part of the cross-country differences in public policies, political instability and other economic factors associated with long-run growth and by evaluating the indirect effect of ethnic diversity on public policy choices that in turn influence long-run growth rates. According to Aker et al. (2016: 193), this phenomenon generates internal borders and market segmentation.

In light of this, we claim that there are two factors to be underlined about Niger: on the one hand the multiplicity of ethnic groups within it (Aker et al. 2016: 193), on the other the tendency towards nomadism that characterizes most of the African territory (Arecchi, Delisse 1999).

Today Niger State has been divided into a total of seven provinces thus; Tahoua, Agadez, Diffa, Zinder, Maradi, Tillaberi and Dosso whereas Niamey is considered as a Department (Idrissa, Decalo 1992).

Postcolonial borders not only assigned parts of the Sahel and Sahara regions to independent states, such as Mali, Niger, Libya, Chad and Sudan, but they also divided tribes, clans and ethnicities. In the case of Niger, several groups constituted a minority in the newly created state, which politically and economicly disadvantaged them, especially those located on the edge of the desert. The newly independent government has generally paid little attention to the social and economic development of the north and the communities residing there (Reitano, Shaw 2015).

However, the other factor is produced by the climate and geography of the territory. Niger, dominated by the strong presence of the desert in the northern area, has always forced an internal organization of its nomadic population, to ensure commercial transport between the northern highlands and the areas below the desert region. Over the centuries, the

Fig. 01.03
ertic landscape with
baobab tree,
uré Giraffe Reserve,
labéri region, Niger
©AT 2017

Fig. 01.04
The dramatic problem of plastic waste damages entire environmental systems and human health, Dosso Village, Niger
©LS 2017

Nigerien territory has represented an important crossroads for commercial exchanges in the trans-Saharan area[7] (Salifou 2002). With the advent of the modern era and the development of new commercial exchange and transportation systems, the travelling vocation of the Nigerien population is disappearing, causing a phenomenon of "sedentarization" of the nomadic ethnic groups (Arecchi, Delisse 1999). In order to face the climatic challenges of the region – such as changes in rainfall, cyclical drought and growing desertification, pastoral and sedentary communities in the Sahara and Sahel – have developed various forms of subsistence. Internal and cross-border migration between communities across the region has served to dampen the worst shocks to people's livelihoods (Tinti, Reitano 2016; Brachet 2012).

The human need to "survive" in a desert territory like Niger has contributed to a continuous flow of exchanges. In essence, it is precisely this crossroads of ethnic groups, cultures, religions and trades over the centuries until today, which generates Niger as a dynamic territory.

Dynamic landscape. Vernacular geo-climatic features

A peculiarity of Africa is that "the landscape is not stable, but transformed by the seasons, by the weather and by the activities of the villagers" (Loflin 1998). In line with this statement the State of Niger, a landlocked country, is grafted, like a precious stone, into the heart of the Sahara Desert. Seemingly monotonous and arid, the Nigerien territory is exceptionally dynamic: exploring the country from North to South, the presence of a variety of landscapes reveals that it has given life over the centuries to a collection of biodiversity. According to the climatic characteristics, the land reflected in the cultural components of the ethnic groups that still existed.

With the aim of defining a coherent project that is the spokesman of the vernacular richness of the territory of Niger, it is, therefore, necessary to understand the dynamism of these natural landscapes, reading heir specific geographical-climatic characteristics and learning the dynamic bonds that they generate.

With the word "dynamism" (Pickett, White 1985; Di Castri et al. 1988), we claim to express the vigorous complexity of the environmental system of the desert. Dynamism means the combination of all the elements that composed it: the dynamism of its natural elements (sand, wind, vegetation, oasis, river, ancient basin valleys) the dynamism of its inhabitants (humans, spirituals and animals) the dynamism of the space of living and the rhythm of life. Everything echoes, everything listens, and the result is an overall balance.

As we already mentioned, Niger is one of the largest and hottest inland countries in West Africa. Its land which represents the traditional gateway between North and sub-Saharan Africa, is almost arid and two-thirds of it is covered by the Sahara Desert.

However, the territory is only apparently monotonous in its features: a vast plateau, whose most typical stone is the laterite,[8] gradually rises from the South-West to the North-East, with an average elevation of 500 m, and it is "shaped" according to the rainfalls. This crucial element has the power to condition the fragile balance in an arid landscape. The presence of rainfall is rarefied in the North and proceeds to rise in the South of the country.

This trend, combined with the geographical settlement, contributes to a natural division of the territory into three different zones (CILSS 2016: 157–159; Bernus et al. 1980):

- the desert zone in the North that accounts for two-thirds of Niger territory, with a dry and hot climate, including the Aïr region, characterized by nomad cultures;
- the intermediate Sahelian zone in the centre, characterized by semi-nomadic sheep farming;
- the South zone, the most fertile region, situated in the South of the state, next to the Niger Delta, which allows the cultivation of most of the resources of the state thanks to the more humid climate regime, and which, therefore, welcomes the majority of the population.

The geography can be defined as dynamic thanks to the climate, that plays a significant role in the morphological conformation of a territory, and it can be considered the activating paradigm of geography. The main climatic features of Niger are represented by extreme conditions managed by the sub-Saharan climate in which it is located: drought and humidity, heavy rain, cold winds and intense heat.

The characteristics of the country are shaped by the changing seasons: the atmospheric circulation at altitude, which determines the orographic condition of the territory, allows the alternation of three main seasons during the year, with a drop in temperature and rainfall affecting all three climatic regions (Bernus et al. 1980: 14; Mester de Parajd 1988: 10–12):

- the dry or cool season (or winter), which is affected by the subtropical climatic conditions, covers the period from October/November until April; temperatures are around 30°C and the least warm months of the year are December and February
- the hot season (or summer) conditioned by equatorial and

southern influences, which runs from March to May with constant temperatures around 40°C

• the rainy season (or wintering season), which occurs mostly in southern Niger between June and October. In fact, at the latitude of Agadez (in the North), no rainy period happens today, due to global warming.

A significant characteristic due to the three seasons is the rotation of two masses of hot air which differ in their humidity. This phenomenon has a massive impact on living conditions and has contributed to the generation of various types of housing within the country to protect the inhabitants from the environmental conditions. Indeed, during the winter season, in January and February, the high subtropical pressures that arise come from the Sahara cause the circulation of an arid continental wind, called the "harmattan", blowing in a south-westerly direction towards the Equator, sometimes causing sandstorms (Prussin 1974). Typically, the wind desiccates, dries and creates dust, upsetting living conditions on the southern fringe of the desert (Mester de Parajd 1988: 12–14).

From April to May, the low pressures particularly affected by the solar supply that the Sahara receives allow the winds of the southern hemisphere, which blow from the Atlantic to the Equator, to penetrate: the "monsoons" are a humid-laden air mass that, when it meets the dry harmattan, causes a violent crash that marks the beginning of the rainy season (Bernus et al. 1980).

The Earth is not a stable element but is transformed by the seasons, the rainfall and human activity. A fascinating characteristic of African territories, rural or urbanized, is that they conform and change according to the changing factors. Following the temporal sequence, the landscape alters its structure: nature humidifies with the rain, makes crops grow, or dries and fragments the soil. The life of the inhabitants is harmonized by the seasons. Even the scenery of the route changes, due to the absence of pavemented roads. All human paths created during the dry season, disappear during the rainfalls, and new routes must be designed, according to the dynamic accessibility of the territory.

The practice of walking is closely linked to the life of the Niger population, as we can extend the concept to all humans living in contexts without infrastructure and road connections, where walking means surviving. All the practices of daily life include distances to cover by walking: to collect the water from the well, reach the school or the place for praying or cultivate the fields. In the African tradition, therefore, the journey takes advantage of a spiritual charge that

Fig. 01.05
Gaya Village, Niger
©LS 2017

Fig. 01.06
Railway Niamey-Cotonou, Dosso region
©AT 2016

brings humans to meet their destiny: walking represents a moment of spirituality, contemplation, fatigue but the achievement of goal. It is precisely during the journey that humans mark their path on the earth, and the soil will carry their footprints. "The earth must first exist as a mental concept. Then you must sing it. Only then can you say it exists" (Chatwin 1987 [1995]: 27).

As the pathways are in an inclusive relationship with the climate mutations, also living space, conceived as a living being, changes according to the seasons. Generally, the typical house in Niger is built with banco[9] – raw earth – that represents the most common and affordable traditional material. However, during the rainy season it usually gets consumed and deformed, and every year the house has to be "reshaped" or rebuilt. Indeed, the climate regime not only affects the life cycle dwellings but also generates a sequence of spaces that make it possible for the inhabitants to live in notably harsh environmental conditions, such as the high temperature during the summer. During the hot season, the primary necessity is the protection from the hot temperature, and ventilation ensures adequate air circulation in these climatic conditions. The traditional banco houses, as evidenced by the few ancient buildings[10] that survive today, are built with vaults, and through a system of fences seek to remove direct sunlight from walls, while small holes create air convection.

As we argued, in the African culture there is a strong connection between nature and humans, and the land is conceived by them as an integral part of their everyday life, living experiences and spiritual life. (Kultermann 1963; Arecchi, Delisse 1999).

The climatic factors have strongly influenced the establishment of people in several places, and architecture has been shaped according to specific conditions in response to human needs. The climate has greatly influenced the environment and the life of its inhabitants, especially in Niger, which is one of the hottest and driest regions on the globe.

The partition of the Niger landscape into three types is the direct consequence of the climatic character, which conditions the living spaces, and consequently the culture and spiritual vision of the population. Hence, understanding the dynamism of the desert is necessary to follow the architectural answers that humans have shaped over the years in response to the combination of different environmental factors and a consequent life practice (Mester de Parajd 1988).

It is crucial to underline that the variety of architectural types (e.g. villages, houses, granaries) are a representation of the traditions of local ethnic groups. However, they constitute

Fig. 01.07
Dosso Landascape
Niger,
©LS 2017

Fig. 01.08
Rice field in Niger River, Gaya region, Niger
©LS 2017

a national and world heritage that is gradually disappearing, replaced by "modern" construction models and new materials. Local techniques are no longer used, thus leading to the progressive abandonment of traditional life systems.

According to some studies on the vernacular architecture of Niger (Houben, Guillaud 2006; Mester de Parajd 1988; Bourdier, Minh-ha 2005; Abbado et al. 2015; Baudouï et al. 2003; Schutyser et al. 2003; Morris 2005), and research surveys carried out by the authors, it is possible to highlight the close relationship between the geo-climatic characteristics of the territory and the morphological variety of local architecture, rediscovering the dynamism of the desert in the concept of living space.

 # 1. Landscapes of the Sahara Desert | habitat nomade |

The north area of Niger is mostly covered by the Sahara Desert and is intersected by numerous depressions, dominated by arid highlands and local reliefs. The highlands of the North are cut by valleys (*kori*) of the Aïr Massif, which is an extension of the Ahaggar, Mountains of Algeria (Thorp 1969).

The origin of the name "Sah'ra" is mentioned for the first time by the Arabic writer Ibn-el-Hakem meaning "void", and describes a wasteland being permanently short of water and hostile for human existence (Schulz et al. 2009: 64).

The expanses of Sahara Desert dunes occupy most of the territory of Niger (about 65 per cent), characterized by an almost total scarcity of water, with substantial thermal variations, rare vegetation and steppes, large gravel and stony plains. This zone corresponds to the districts of Agadez and Diffa, where the Sahel gives way to the Sahara, and extends to the border with Algeria and Libya. The extension of this biotype drastically affects the population density of Niger, which is pretty low: 12 ab/m^2, as 94 per cent of the inhabitants occupy only 35 per cent of the whole state (World Population Review 2020).

Among the most striking natural features of the Sahara Desert are the Aïr Mountains, which rise from the sands of central Niger. These mountains, which cover an area of 72,000 km^2, have some of the most magnificent landscapes in West Africa (CILSS 2016: 11). Their origins are volcanic and they have evolved with high peaks (Mont Bagzane 2022 m). Much of the massif is supported by Precambrian yam and metamorphic rocks, characteristic of other large massifs found in North Africa (Lowman 1968; Thorp 1969). The mountain masses are round and prominent, over 45 km wide. The imposing dunes of the Ténéré Desert are stacked against the escarpments of the eastern side of the mountains thus combining two

Fig. 01.09
Desert landscape in Dosso region, Niger
©LS 2017

Fig. 01.10
Fields in Tillabéri region, Niger
©AT 2016

extraordinarily beautiful landscapes. Here, humans have lived here for thousands of years. Neolithic art finds are present in the valleys. The paintings represent a lush savannah, full of many of the large mammals (CILSS 2016: 11).

Among these desert mountains, there are oases and natural springs, as in the case of the village of Timia, which has date palms and green gardens. Traditionally, the Aïr lands have been inhabited by the Tuareg people. These are nomadic shepherds who live in a large area of central and western Sahara. Many have settled in small towns such as Timia and Iferouane, or Agadez, the gateway to northern Niger. Addressing a climate and a desert territory implies life solutions that must completely adapt to the context. For this reason, nomadism was established as a housing practice in this specific environment. The architectural response and the space to live has been shaped by these needs (Arecchi 1991: 36–42). The traditional Tuareg's habitat consists of a light and elastic structure, easy to disassemble and transport, which must face the arid and torrid climate, the presence of wind and sandstorms. The cover is a tent composed of a set of goat skins sewn together, lying on a wooden structure, anchored to the ground (Taton 1984: 9–12). The shape of the spaces varies, but they mostly are rectangular. Due to the high temperatures, constant ventilation is crucial, therefore the tent is open on all sides: a dense weave of plant fibres and wool acts as a visual separation from the outside (Mester de Parajd 1988: 22).

Only in the Aïr region, where the terrain is rocky, there are rare cases of stone houses; in the ancient city of Assodé, now in ruins, dwelling models were found which included walls made up of 50 per cent of rough stones and earth (Mester de Parajd 1988: 53).

 # 2. The Sahel region | pastoral habitat |

The Sahel[11] region, from Arabic "Sahil", "edge of the desert" is a broad semi-arid belt, extending from the Atlantic Ocean to Sudan (and to the Red Sea), averaging about 350 km wide. Niger is one of the countries included in this climatic belt, a transition zone between the desert and tropical West Africa, located in the central area of the state. This agro-pastoral region is drier and greener than the Saharan one, it has an annual rainfall of between 150 and 600 mm (CILSS 2016: 7), which varies from year to year, and a dry season that lasts 8 to 9 months. The vegetation is generally characterized by steppes, savannas, the typical tiger bush[12] and short grass (Aristida and Cenchrus), mixed with woody plants and thorny trees from the genus *Acacia*.

However, usually, the number of woody plant species is relatively low.

Fig. 01.11
The well of Kafi village,
Gaya region, Niger
©LS 2017

Fig. 01.12
Around Kafi village, Dosso region, Niger
©LS 2017

The Sahelian vegetation present is the typical result of long-term human and animal presence. Where there is ample grass cover annual grass fires often sweep across the desert's landscape (CILSS 2016: 7).

In this Sahelian region are located very significant city centres, which are cultural representatives of the country, such as Tahoua, Maradi and Zinder. The Maradi-Zinder region constitutes the largest agricultural area in Niger. The level of bio productivity is higher in the Central-Southern areas, where there are many trees, thanks to the less arid climate, which allows a richer natural regeneration. The South-West region, the valley of the ancient river basins of the dallol, is in sharp contrast to the steppe and tiger bush area present in the surrounding regions, thanks to its wetland nature, which allows the cultivation of the land and human settlements.

Particularly interesting is the architectural response to the climate of the Sahel, consisting of banco houses and straw huts. Originating in the Haussa countries, and then also brought to other ethnic groups, such as Djerma or Kanouri, this type of construction is spread over the whole Niger territory. The choice of this construction technique derives from the internal well-being of the banco structures which promotes excellent thermal comfort during the dry seasons, managing to trap the freshness of the night and then return it during the day. Instead, the problem with this material is during the wet season. For this reason, in the South, this type of housing solution is replaced with other materials (Mester de Parajd 1988, Taton 1984: 15–28).

However, this type of dwelling only ensures some durability if it is maintained seasonally after the rains which deform the casing. For this reason, especially in urban centres, the use of concrete bricks and other materials that are considered long-lasting over time, is taking over.

 # 3. Sudanian region and semi-humid area | sedentary habitat |

The Sudanian region and semi-humid area covers about 1,600,000 km^2. In this area the development of vegetation and fauna is much more prosperous thanks to a milder climate and a more generous rainy season and the presence of the Niger River which makes it possible to cultivate rice and cereals. The periodic overflow of the Niger River, virtually the only river in the country, guarantees the cultivation of crops in this region (Idrissa, Decalo 1992; Aloko et al. 2010), but also creates correlated issues in terms of human and natural disasters.[13]

The southern Sudanese plateau, which covers an area about 1450 km in length, is subdivided Southwest into the Djerma-Ganda region, a vast sandy valley, where the dallol descend from the Aïr's highlands; a central area,

Fig. 01.13
pastor of a village in
Dosso region, Niger
©LS 2017

Fig. 01.14
Gaya region, Niger
© LS 2017

consisting of the Adar Doutchi Massif and the Majia area, which includes the area of the gulbi, dried-up valleys of former tributaries of the Sokoto River of Nigeria. Even in the southernmost area to the east, ancient paths of water streams can be seen within the sandy expanse of the Manga region.

In the Sudan region, where the rains are more significant than in the northern desert and central areas, there are types of homes that follow the passage of the seasons. Typical of the Djerma and Shongai populations, located in the South and South-East of the country, is the hut with vegetable mats used as a covering. The co-construction of the house begins in the dry season, when a wooden structure, which can be rectangular or circular according to tradition, holds up the cover of plant texture. After the end of the rainy season, the construction of the dwelling is completed with a layer of banco, which will be remodelled at the end of the next rainy season (Mester de Parajd 1988). The rhythm of nature dictates the type of architecture.

Over time, architectural solutions have been formulated that have tried to adapt and propose solutions appropriate for the climate. Notably, in Niger, where part of the year the climate is hot and dry and the other part, for three or four months, is warm and humid, it becomes problematic to combine these two realities. However, we can see how the profound knowledge of the geographical context has been able to establish brilliant technical and construction solutions over time using local materials, moulded to meet spatial and living needs.

Ecotones of the landscape: Dallol as dynamic paradigms

The "ecotone" is considered to be a tension zone, an interchange of energy, of matter and organisms between two different ecotypes (*Merriam-Webster 2021*): an area in which two orders meet and exchange components (Farina 1998: 94), a collision between two homogeneous systems that through their union generate complexity, thus development. The greatness of the ecotone, defined as an environmental vector, lies in creating interactive links, synthesis and generation of new models.

The territory of Niger is considered dynamic for the interaction of environmental phenomena that have a direct consequence on the morphology of space for human existence and finds in its sub-Saharan nature the presence of the activating ecotones: the Dallol.[14]

The ecotone represents the interaction and transition tool between the environment and the needs of life. We argue that it is a fundamental methodology of study and analysis

Fig. 01.15
Humid land, Dallol
Gaya region, Niger
©LS 2017

Fig. 01.16
Habitat in humid land, Gaya region, Niger
©LS 2017

for efficient design practice, rooted in vernacular knowledge and projected into new future evolutions.

The dynamism of the Nigerien territory describes a space of interaction between "ecosystems" of different types: the coexistence of various adjacent ecosystems, and the union and synergy within these elements, which can be considered part of a single element – the landscape – is possible through the reading of the ecotones. As the subject of numerous studies in the field of ecology, the ecotone[15] is described by the first ecologist Clements (1905) as a tension zone and refers to a transition area between different adjacent ecological communities that are structurally different (Sauer 1995). This environmental vector is considered to be a dynamic[16] interface responsible for the constant balance of interaction between contiguous ecosystems. The ecotone may be considered a true habitat, but also a frontier habitat, where overlap encourages species diversity (Harris 1988).

According to Farina (1998), ecotones are therefore critical structures for the landscape and its functioning: "nutrients, water, spore, seeds and animals move across these structures and play an important role as indicators of climate change caused by increases in CO_2 and air pollution" (Farina 1998: 94). The spatial heterogeneity defines a system with complexity and variability (Li, Reynolds 1994). The variations reflect the rate of change, in processes and functions, as heterogeneity is the main characteristic of each landscape (Hansen et al. 1992) and may be perceived at any scale of investigation. A clear definition of heterogeneity is fundamental for approaching and understanding the different environments and defining the approaches of the project.

The researched Nigerien Ecotones include lots of different geographic typologies, such as groundwater (Décamps et al. 1988) and surface water (Fraser, Williams 1998). These two ecotones represent precisely a possible tool for reading and understanding Nigerien territory.

In this context, the "Dallol" is considered to be a specific and vernacular ecotone, as it revitalizes the land by means of ancient river basins, making a transition zone between the desert and wetlands, encouraging the creation of settlements and habitat areas, due to the richness of the soil and biodiversity.

Dallols, the ecotone of Niger

There are two major hydrological systems in the desert territory of Niger: in the South-West the basin of the Niger River, in the South-East the lake Tchad. These two sources of water are now active and exploited for commercial purposes and constitute the "simple hydrological system" of the southern part of the state. This system is entirely dissimilar from the ancient hydrological one which characterized the northern part of the country's desert over the centuries and that allowed people to settle in the Aïr regions about 500–600 years ago becoming shepherds and farmers. These two hydrological systems can be connected by "dynamic" elements of the territory, which link the Aïr region to the Niger River: the dallol.

Dallol is a word in Dgerma language, which means "ancient river basin", and represents a vital element for human living in arid desert zones: dallol, or fossil valleys, are underground waterways, where the water flows almost flush with the ground level. During the rainy season, the water moves superficially, while during the rest of the year it creates a humid area, retaining the water underground, and leaving belts of water on the surface called "mares" (Boudet 1969). The fossil watercourses are relics of a former pluvial age. The principal system runs south-westward in an arc from the Aïr Mountains to join the Niger River in the Dallol Bosso, while the Dallol Maouri, on the East, and Dallol Foga represent a subsidiary system (Rochette et al. 1966). The Dallol Bosso is 775 km long but is discontinuous over stretches of 30–125 km in its middle course. The entire ecosystem probably comprises 750,000 ha (Bernacsek et al. 1992).

Due to this inestimable orographic character, the dallols represents an attraction for human settlements and agriculture. Over time these natural areas in an invariably homogeneous environment have allowed the development of the types of flora and fauna which have made human settlements possible. The dallol banks are mainly dotted with unique tree species such as the *ficus gnaphalocarpa*, from the *Moraceae* family, the *Hyphaene Thebaïca* and the *Borassus Aethiopum*, of palm types. There are also some relevant herbaceous species called *Cypéracées*, essential for the cultivation of potatoes and sugar cane, much of which is practised in these areas (Boudet 1969: 306).

Therefore, the dallol is a fundamental orographic element for the morphology of settlements in the Niger territory: it designs the panorama and defines itself as an essential element for reading the landscape. The presence of dallols within the desert territory of Niger acts as mediation, a transit zone, exchange section, meeting point between different environmental systems, fauna, flora and ethnic cultures. According to Farina's (1995; 1998) definition, we decode dallol as the main ecotone[17] for the Niger territory. It also represents an element where the population transit through these areas in search of places to settle taking advantage of the humidity of the soil, and where plant and animal species that are thoroughly different from the rest of the region come to life. Therefore, it is a dynamic ecotone because its contour changes seasonally due to the flow of water that moulds its shape.

Research has been carried out by the authors into this topic in Dosso department, located in the south part of the country, where the convergence of three Dallols (Bosso, Maouri and Foga) towards the Niger river basin is particularly significant. The study (Spadari 2017) aimed to investigate the interaction between the dallols' ecotone and the consequences in specific local architecture, as the ecotone is the most characteristic element of the territory and strongly influences the settlement morphology of humans within the territory.

The context reading aims to summarize the dynamic elements of the territory, marking a system of formal and informal axes. The formal axes are the infrastructural axes represented by the only two national roads that connect the Dosso department to the capital Niamey. The road was traced for economic reasons since the Dosso department is the backbone of the country's fragile commercial and economic development. The primary exchange centres are located along this route, and this line becomes the magnetic pole for the whole territory. Hence, with the advent of the two

e founding cities can be represented within an orthogonal grid system with regular distances

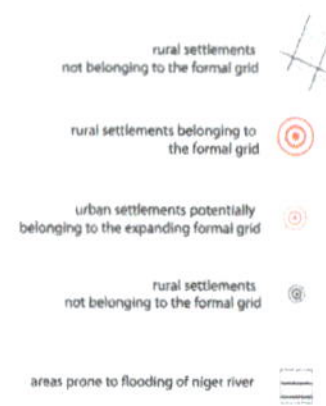

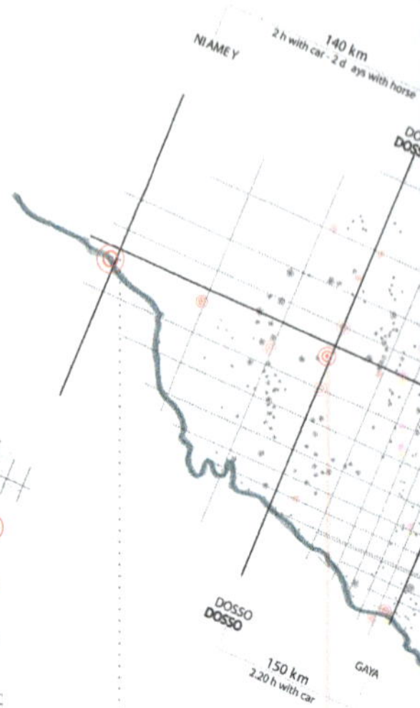

commercial network_ FROMAL CONN

n routes of the Department of Dosso are the National Roads ?1 Niamey-Diffa e N7 Dosso-Gaya) that divide the territory according to the two "commercial" axes

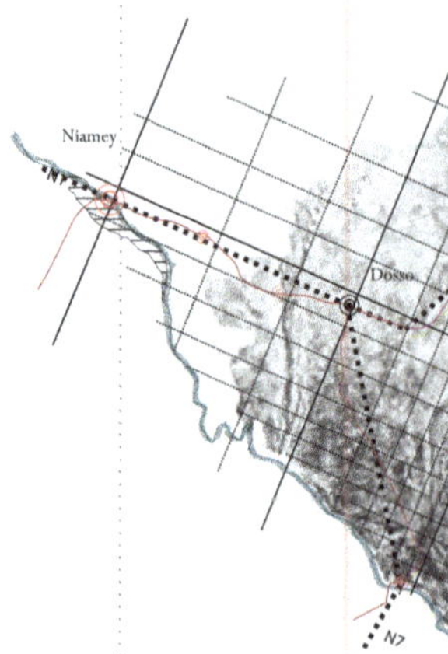

urban expansion_ RIGID MORPHOLO

The commercial axes are configured as superimposed infrastructures, ich do not establish any relationship with the morphology of the territory

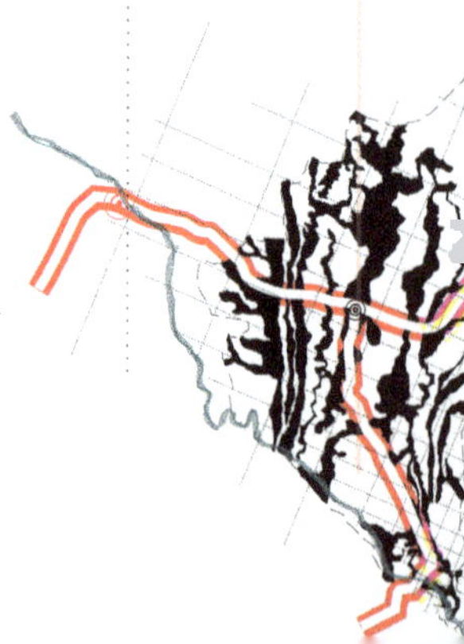

Rural settlement_ *Informal Grid*

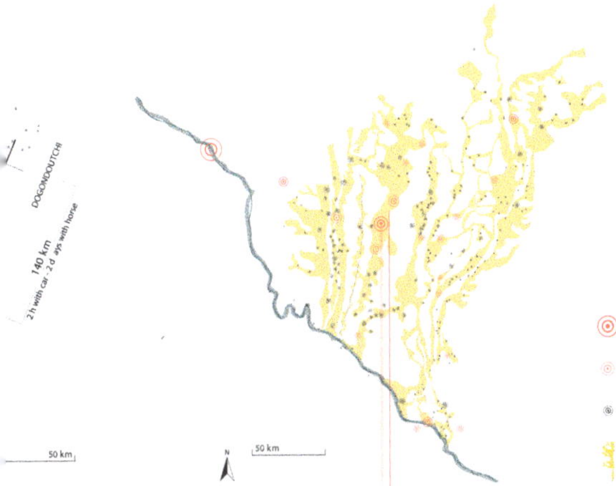

The morphology of the territory corresponds
to the morphology of the living space

landscape morfology _ INFORMAL CONNECTIONS

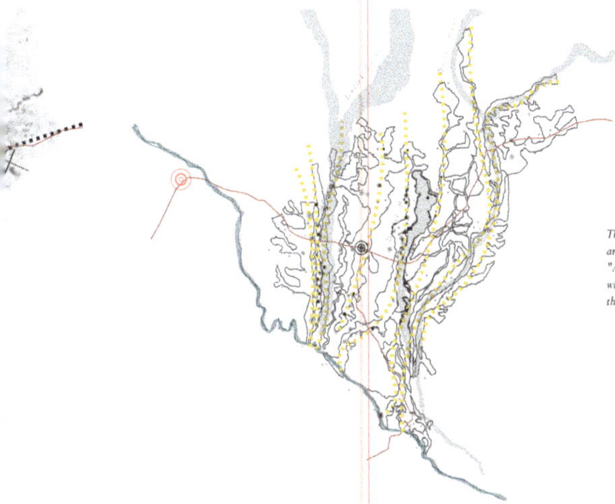

The strongest population concentrations
are found along the "dallol"
"Ancient Basin" in the Dgerma language,
which represent southern orographic axes
that converge to the Niger River.

Fig. 01.17
Study of
formal-informal grid of
settlements in Dosso
region, Niger
©LS 2017

landscape eco-tonality_ DYNAMIC MORPHOLOGY
of eco-tone and HABITAT SPACE

The morphology of living space
is directly related to the conformation
of the territory designed by
numerous ecotones, the dallol ,
which converge in the Niger River

national roadways, many villages have changed their location, moving closer to the road networks and trying to take advantage of the possible economic benefits which these paths can bring. The natural axis system, in contrast to the formal one, is constituted by dallol ecotone, which contributed to the informal planning of the entire local rural settlement. The Bosso, Foga and Mouri Dallols are the three natural guides, identifiable as wetlands, which have allowed humans to gain asylum from the arid conditions of the desert (Rochette et al. 1966). The proximity of water and more luxuriant flora and fauna led people to settle along the ancient river basins, modelling the rural settlement systems in the region. According to the maps, it is possible to observe the constellation of villages gathered along the dallol ecotones, which can be examined as a design tool, being interpreted as green infrastructure.

Openings

A coherent development of the territory must be able to read the tracks that the history of the natural and human landscape has generated. In this context, development processes, migrations, climate changes (e.g. desertification, deforestation) shape the geomorphology of the territory. Then, the ecotones become a consistent

Fig. 01.18
Dynamic geography, Dosso region, Niger

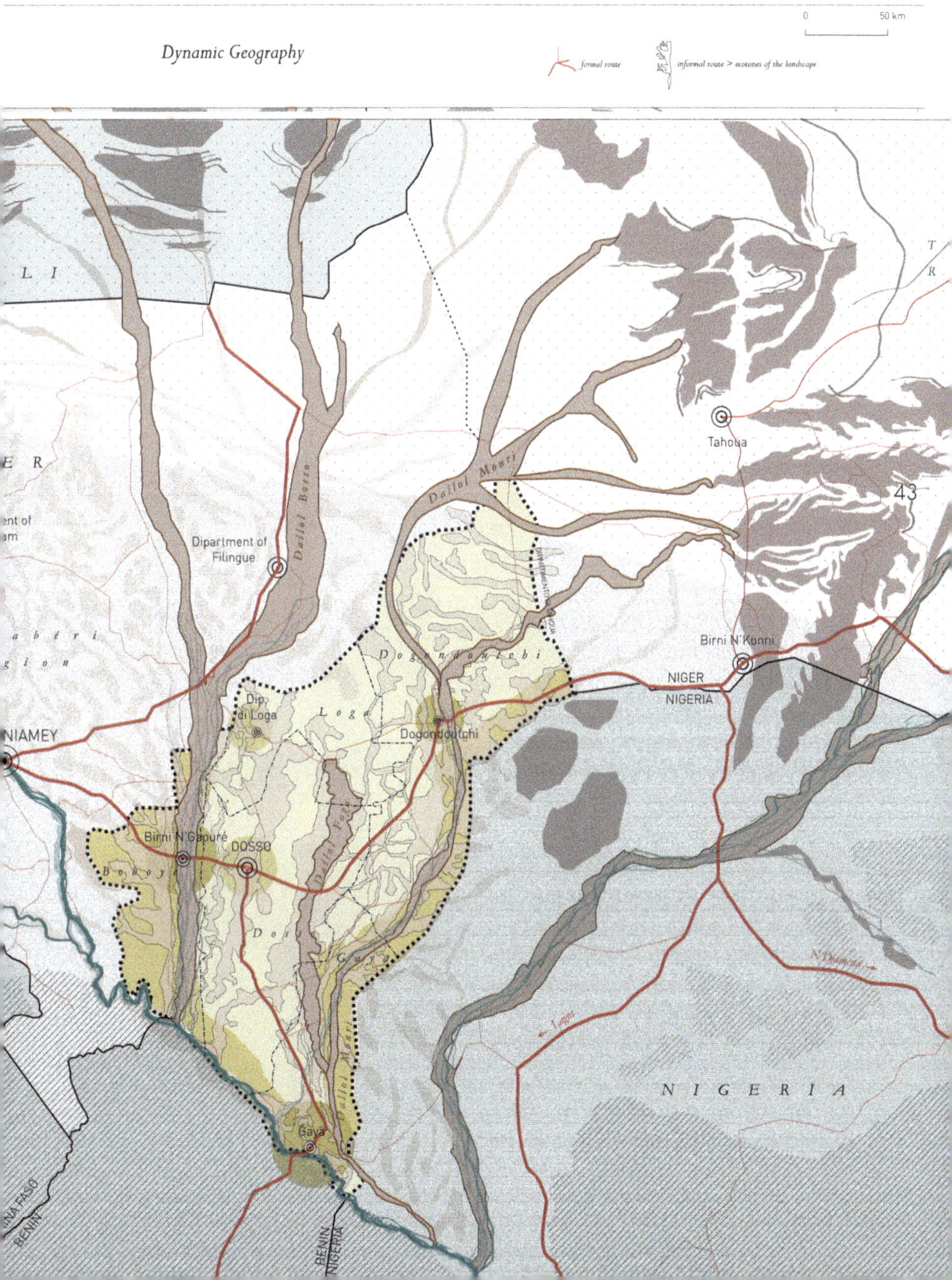

element, which has influenced human settlements over the centuries.

Today the Niger context is plagued by conditions of poverty that have fragmented and ruined its cultural values, and it is precisely in the possible attention to and regeneration of local ecotones that it may be possible to reactivate the territory.

In the present research, the ancient fossil valleys of dallols are considered to be the axes of a natural regulatory framework, which has governed human settlement in the area. The informal approach of choosing a place to live by finding a compromise between the natural landscape and its habitability made sure that the dallol ecotone led to the spontaneous aggregation of living spaces.

Therefore, in design philosophy, the landscape reading, recognized as a condensation of layers and natural and anthropic memories, becomes crucial. The ecotone is considered to be an element of mediation between mnemonic knowledge and innovative thrusts and establishes itself as a virtuous model of development. Studying the Nigerien landscape becomes the framework for setting a new building model, which sinks into the collective memory of vernacular architecture techniques.

These concepts aim to define a better quality of living and respect for the territory in which we live, according to mediation between the

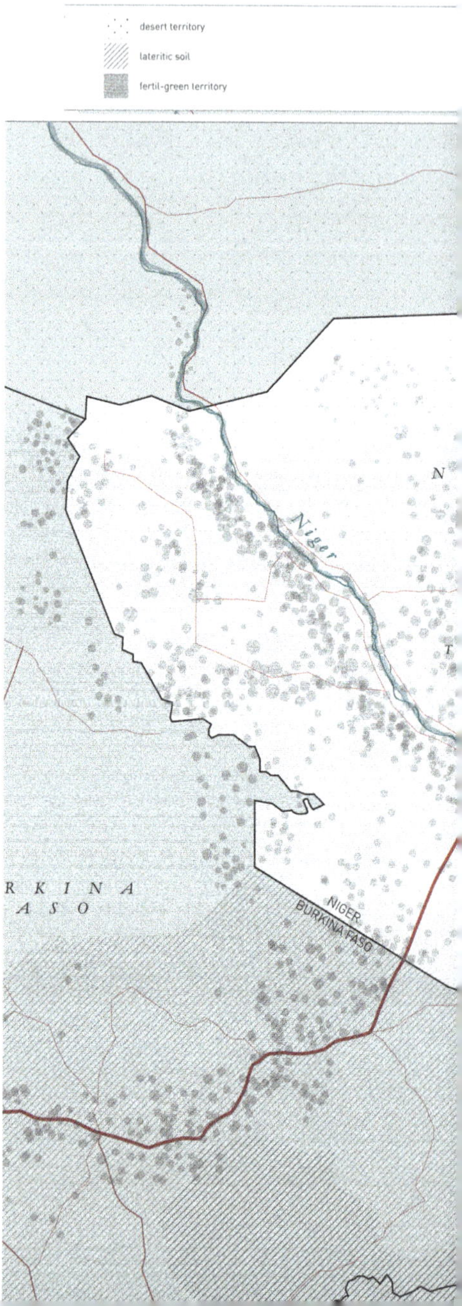

Fig. 01.19
Morphology of Settlements
Dosso region, Niger
©LS 2017

rediscovery of lost knowledge, new technologies and protection of the natural environment.

>>> Notes

1 Niger Delta is characterized by a delicate ecosystem in which local populations live in balance with nature. The land consists of varied ecosystems, such as mangrove swamps, freshwater swamps, rain forest, and one of the largest and most important wetlands and marine ecosystems in the world. When the first oil fields were discovered in 1956, the oil companies (Shell controls about half of the total crude oil, Total, Chevron and the Italian Eni) colonized the territory, supported by corrupt military governments, which have sold off their country's natural resources in exchange for illicit profits. Today the Niger Delta, due to oil pollution, is characterized by contaminated streams and rivers, forest destruction and general loss of biodiversity. This situation affects the livelihood of indigenous people who depend on the ecosystem services for survival, increasing poverty and displacement of people. Although the oil industry has contributed to the growth and development of the country in a way, it has rendered the Niger Delta region one of the five most severely petroleum damaged ecosystems in the world (Kadafa 2012, Allen 2012).

2 https://www.statista.com/statistics/265195/oil-production-in-nigeria-in-barrels-per-day/ [access on 14th March 2019].

3 According to UNHCR, the humanitarian and political crisis in the Sahel region has reached unprecedented levels, and this increases the migrations in western and central Africa. The worsening security situation in Mali, the attacks and kidnappings in Niger, the violence and attacks on civilians in Burkina Faso, the security instability in north eastern Nigeria are causing large-scale forced displacements within and beyond borders. Documenting the scale and nature of risks and abuses along travel routes in West and Central Africa is challenging. The refugees crossing western and northern Africa often face a dangerous journey that has included passing through areas of armed conflict, crossing deserts and, for some, being held for ransom and tortured, or trafficked for sexual or labour exploitation (UNHCR 2019).

4 "Between 1990 and 2018, Niger's HDI value increased from 0.213 to 0.377, an increase of 76.8 per cent. Between these year, Niger's life expectancy at birth increased by 18.5 years, mean years of schooling increased by 1.3 years and expected years of schooling increased by 4.4 years. Niger's GNI per capita increased by about 4.5 per cent between 1990 and 2018" (Human Development Report 2019: 3).

5 "The most recent survey data that were publicly available for Niger's MPI estimation refer to 2012. In Niger, 90.5 percent of the population (19,431 thousand people) are multidimensionally poor while an additional 5.1 percent are classified as vulnerable to multidimensional poverty (1,094 thousand people). The breadth of deprivation (intensity) in Niger, which is the average deprivation score experienced by people in multidimensional poverty, is 65.2 percent" (Human Development Report 2019: 6).

6 As a result of the ongoing desertification since 1965, the population has been pushed

The conceptual apparatus of tradition

02

Alisia Tognon

DOI: 10.1201/9781003157632-02

The importance of talking about vernacular architecture is due to the extensive and complex relationship that regards the teaching, application and significance of this concept within the modern context, and the possibility of responding to the contemporary environmental issues. This observation is directed to consider why and how it can be helpful to investigate the sense of vernacular architecture in the twenty-first century.

We argued that there are valuable lessons to be learned from the traditional knowledge, skills and competences of the world's vernacular builders. Furthermore, the analysis of the values of vernacular traditions shows that the importance of its study is explicit in various fields: housing, conservation, sustainable development and even disaster management.

Supporting its centrality in architectural studies, we aim to demonstrate how knowing the vernacular becomes the core for providing environments built in the future and peculiar contexts. Hence, this approach will allow the active implementation of vernacular know-how in a contemporary framework. The vernacular importance is applicable all over the world, and even more in these contexts where the environmental conditions are involved. Nevertheless, according to Asquith and Vellinga (2006), in a time of rapid technological developments and globalization, vernacular architecture still occupies a marginal position. Even today, architectural education, research and practice largely ignore the results, the experience of the vernacular and the skills of builders. However, faced with the persistent denial of the importance of vernacular architecture, questions arise about its function and meaning in the twenty-first century.

The subject of African architecture was – and indeed still is among many – not considered worthy of recognition (Prussin 1974). In the past, when referring to African architecture, its centrality has long been diminished, generalizing it with the term "mud hut".[1] The architectural culture has been downgraded and limited to simple rural buildings, considered poor and underdeveloped. However, as Arecchi

(1991) argues, it is possible to find the roots of a rich tradition of living forms, which has produced its monuments, and has found itself limited to making standardized rather than quality spaces (Arecchi 1991: 14). Today the demand for modernization according to a misleading westernization has meant that we tend to forget the traditional models of living. In the meantime, western materials and construction techniques are adopted, although they constitute an incorrect response to a differrent climate where they are wrongly used.

The relevance of the study of vernacular knowledge is emphasized by reference to current global concerns over sustainability and cultural identity which continue to question the processes of modernization and globalization (Asquith, Vellinga 2006: 10). This research field implies a holistic, integrated and, above all, critical approach that integrates the exploration of materials, technologies and environmental performance, adding an examination about the social, political and economic aspects (Vellinga 2015).

The vernacular knowledge must be integrated with modern knowledge to allow the development of contextually, culturally and ecologically appropriate settlements, intending to create meaningful places.

Therefore, the reflections developed in this chapter start from foundational questions, which arise working in a context such as the African one. How could the nature of local culture and traditions and their structure influence the adaptation rate of external architectural innovations? How does the cultural and social status of the local architectural elements or types determine the import rate of innovations? What are the conditions that can lead to the friction and decay of an indigenous vernacular architecture?

Vernacular architecture and critical regionalism

"In using the generally accepted phrase 'vernacular architecture', I am embracing all the types of building made by people in tribal, folk, peasant and popular societies where an architect, or specialist designer, is not employed" (Oliver 2006: 4).

Today, speaking about vernacular architecture in a context such as Africa looks necessary due to environmental needs. However, it is even more important to understand the meaning of a vernacular concept and how this can shift in a practical sense into critical regionalism.

Although it is entirely appropriate that the definitions of vernacular may vary among scholars, the issue needs to be clarified. It has received a variety of meanings, rarely with the same

Fig. 02.01
Tata Somba House,
Natitingou, Benin
© AT 2016

Fig. 02.02
Village near salt lake, Dosso region, Niger
© AT 2016

value (Heath 2003). It is noteworthy that lately the vernacular term, which has generally established itself as a study of past years, has been accused of traditionalism, nostalgia and even ethnocentrism (Wells 1986: 1).

Nevertheless, the bridge between vernacular architecture theory and contemporary regional design has been a central concern when we are working on and researching these kinds of African architectural contexts.

The word "vernacular" is derived from the Latin "vernaculus", meaning "native", so in architecture, the definition means "native science of building". In usage, however, "vernacular" generally refers to the language or dialect of a people, while architecture is given a qualitative status (Oliver 2006: 4).

Wells (1986) argues that the conventional view reported by several authors (Roberts 1985) defines the applied "vernacular" a more comprehensive range of types of buildings – factories, owned houses, farms and schools – as well as entire landscapes – mills, cities, suburbs, agricultural settlements and commercial strips, which have been usefully classified and analysed as vernacular architecture. However, until recently, the term "vernacular" was rarely used and usually only referred to linguistic characters recognizable at the regional level.

Upton (1983) claims that the social sciences in the 1960s and 1970s tended to apply the definition of the "vernacular" to the nineteenth Century, concerning rural domestic architecture built before the industrial revolution and generally, before the Renaissance. Hence, the vernacular was often equated with medieval and post-medieval popular activities. In more recent years, in an attempt to tackle the neglected American architecture of "common man", "vernacular" has become a generic term for anything that is not considered "high style" (Heath 2003). It is relevant to note that almost all academic fields have ventured into the methodology of this field of study: geography, folklore, industrial and historical archaeology, culture and social history, sociology, anthropology, history of art and architecture. However, we can say that in the literature's study, the term "vernacular" has lost a coherent meaning.[2]

As Oliver states "there's something cozy about vernacular architecture" (1984: 17). In the international field of vernacular studies, Paul Oliver[3] has established a milestone in opening the way to the interest in these architectures not yet considered. Furthermore, his contribution goes beyond the field of vernacular architecture studies and today still has particular relevance for our understanding of architecture in general (Vellinga 2014: 3–7).

The publication of the monumental *Encyclopedia of Vernacular Architecture of the World* (EVAW) in 1997 marked a decisive point in the development of the field of international vernacular studies. In a real sense, it helped to validate the subject as an essential field of academic study.

However, the research in this field started in the 1960s with the text written by Rudofsky (1964), which states "Vernacular architecture does not go through fashion cycles. It is nearly immutable, indeed, unimprovable, since it serves its purpose to perfection. As a rule, the origin of indigenous building forms and construction methods is lost in the distant past" (1964: 4). The aim was "to break down our narrow concepts of the art of building by introducing the unfamiliar world of non-pedigreed architecture" (1964: 4). The object became "how much to learn from that kind of architecture before it became an expert's art [because the] untutored builders – in space and time – demonstrate an admirable talent for fitting their buildings into the natural surroundings" (1964: 8). Other pioneers of architectural criticism, such as Rapoport (1969), Alexander (1968), Jencks and Baird (1969) and Norberg Schulz (1971) argued the underlying assumption that architecture includes the total man-built environment and its quality derives from "man's identifying himself with what he builds, using it as a means of self-expression" (Allsopp 1970: 83).

Rapoport (1969) set a new theoretical framework for the study of vernacular architecture. He recognizes that socio-cultural factors determine the form of vernacular buildings, whereas climate, materials, construction and technology are modifying factors. He considered the underlying theoretical constructs that shaped and continue to shape the built environment, including religion, beliefs, customs and socio-cultural forces in general. An exciting aspect of Rapoport's book is his vision, according to which the shape of the house is not simply the result of physical forces or individual causal factors but is the consequence of a whole range of cultural factors seen in broader terms. Based on the work of Mumford (1961), Gould and Kolb (1964) and Redfield (1965), Rapoport claimed that "primitive" buildings were produced by "primitive" societies that had widespread knowledge of everything aside of all to do with elementary technology. The book linked behaviour and form and theorized that the constructed form influences behaviour, not causally but in terms of "coincidences". However, Rapoport denies that "the climate and the need for shelter" have determined the shape of the houses. However, after affirming the supremacy of culture over climate in determining the shape of the house, he stated that the correct response to the climate is a characteristic of primitive and vernacular buildings.

A specific interest in vernacular architecture emerged for the first time in the early 1980s dealing with the interest concerning sustainable architecture. At the Passive and Low Energy Architecture Conference (PLEA) held in Bermuda in 1982, the first studies were presented regarding the environmental performance of vernacular architecture. Subsequently, Hassan Fathy studied the principles of vernacular climatic design in warm and arid countries with Natural Energy and Vernacular Architecture (1986).

However, despite this early start, until the late 1990s, academic interest in the sustainability of vernacular architecture remained relatively limited, when sustainability, in general, emerged as a topic of political, academic and popular interest. Until then, most of the work focused on the relationship between vernacular architecture and the natural environment in rather general terms. However, after the publication of Oliver's book (1984), a significant growth in interest in the sustainability of vernacular architecture is evident, which has led to the proliferation in conferences and the publication of an ever-increasing number of articles, magazines and books (e.g. Asquith, Vellinga 2006; Frey, Bouchain 2010; Weber, Yannas 2014; Correia et al. 2014; Mileto et al. 2015; Sayigh 2019).

In light of this growing interest, today, we prefer to speak about "critical regionalism" which opens up a sustainable approach to architecture. Beginning with Lewis Mumford's writing, the term was coined by Alexander Tzonis and Liane Lefaivre in 1981 in *The Grid in the Pathway*. Later it was taken up by Kenneth Frampton, who elaborated the meaning. As early as the 1980s, Frampton mapped the thematic of "critical regionalism" with the phenomenological aspiration for duality between national culture and civilization. In general, the theory and discourse of "critical regionalism" is a cultural critique of postmodern society and architecture and previous theories of regionalism. It is an approach to architectural production directed to resist a series of physical, cultural and social changes. It is a theory and practice that aims to establish a dialectic between an increasingly globalized civilization and local traditions, however without nostalgia or romanticism.

The Grid and the Pathway has identified critical regionalism as an approach to the design and planning that promotes the ecological, social and intellectual uniqueness, and the environmental diversity of the regions. In the conception of Tzonis and Lefaivre, the critical regionalist approach helps architects to gain the social and political models of progressive modernism from the alienation and distress that characterized the postmodernism of the 1980s.

According to Frampton, "Critical Regionalism necessarily involves a more directly dialectical relation with nature than the more abstract, formal traditions of modern avant-garde architecture allow" (Frampton, 1983: 26). Following Ricoeur's lead, Frampton described critical regionalism's emergence as a self-conscious reaction to the "global modernization [that] continues to undermine, with ever-increasing force, all forms of traditional, agrarian-based, autochthonous culture" (Ricoeur 1965: 276). Frampton affirms the need to create an architectural structure that combines local culture and landscape qualities. The creation of new architecture in the natural environment, where these two elements should be combined in order to create not abstract objects but a concrete symbiosis. The geographical characteristics and cultural heritage will be decisive for the ecology, climate and symbolic aspect of the place. This is creating the "post-form" balance between the natural environment and the cultural heritage that reflects the identity of societies.

In this context, the birth of interest in African architecture is also within the revolution for architectural theories that consider not only the monument worthy of interest. On the contrary, architectural thought has begun to reflect in a broader frame of an anthropogeographic landscape which is the result of man's relationship with the world around him.

It is interesting to analyse what type of literature has arisen in the past years around African Architecture. An in-depth study on that is the eight volumes that make up UNESCO's, *General History of Africa*, first published in 1989. Noteworthy is the general publication about African architecture written by Susan Denyer *African Traditional Architecture* (1978) Other remarkable publications that approach the theme in general terms of understanding the continent are Paul Oliver's *Shelter in Africa* (1971), and *Dwellings: The House across the World* (1987), René Gardi's Indigenous *African Architecture* (1973), Enrico Guidoni's *Primitive Architecture* (1975), Peter Garlake's *The Kingdoms of Africa* (1978), Nnamdi Elleh's *African Architecture: Evolution and Transformation* (1997), and the works written by Labelle Prussin, since the 1970s.

The specific literature on the theme of Nigerien architecture remains limited and French-speaking. It is interesting to note that the period in which the highest number of publications flourished dates back to investigative campaigns between the 1970s and 1980s. Today, there are research centres, such as CISP[4] which is carrying out an inventory of the architectural heritage on earth in the eight regions of Niger, but recent publications on that topic are limited.

One of the common points that we want to highlight in this chapter is that in the general topic of vernacular architecture, as well as in the specific case of Africa, the vernacular builders are anonymous and unconscious[5]. Technical and stylistic science was passed down from generation to generation and was closely linked to collective memory, religion, social identity and ethnicity. Innovation took place slowly, following the changes of the natural rhythms, while the solutions were the outcomes of centuries in which enterprising builders chose (or were forced) to face the various realities of a region. According to the various ambitions of vernacular builders, we tend to approach the vernacular architecture not as an art, but as a tangible demonstration of culture. Vernacular buildings become milestones for information as cultural models, sources of tradition and development of construction techniques. Vernacular architecture, therefore, tends to be treated as an artefact whose importance lies not in style, but its potential as an informer of culture.

What is presented in this chapter is a set of artefacts that are the concrete results of the architectural decisions made by "ordinary people" during "ordinary life". These "vernacular" buildings have produced new and sometimes surprising insights into the cultures they represent.

The need for their knowledge is due to awareness from the environmental point of view of the change in global concern regarding the use of energy and resources (Wines, Jodidio 2000; Cox 2009; Friedman 2012). So, the primary purpose of this work is to evaluate how specific vernacular traditions are ecologically sustainable, going beyond the mere documentation of vernacular forms, materials and functions, which characterize the buildings and places of the Nigerien landscape. A new responsibility for environmental problems pushes inexorably towards a change of trend towards more energy-efficient forms. Besides, the study of vernacular architecture represents the first step in defining new local forms of construction, exploiting the knowledge of how buildings can be expertly designed and promoting cultural conservation and traditional wisdom (Oliver 2003; Rapoport 2005).

So, according to Oliver, we are obliged to repeat the question "why study vernacular architecture?" (Oliver 2006: 5).

The sense of identity: between genius loci and tectonic solutions in Nigerien architecture

From the literature investigation, the vernacular/critical regionalism concept is pointed out as a scientific notion (AlSayyad 2004; Vellinga 2005; Asquith, Vellinga 2006, Maudlin 2010). The vernacular field suffers from lack of theory and an absence of analytical and modelling tools that examine the grades of transformation processes, which interact to produce a change in the vernacular traditions of a region (Al-Qawasmi 2014).

The research claims not to be a slavish return to traditional architecture but rather aims to provide a theoretical framework that brings out a new contemporary "vernacular" of a region, which can evolve, transform or even die. Such an approach will allow us to go beyond the formalist and reductionist definition of the vernacular as a category limited to a historically fixed building typology and help us to examine contemporary architecture as a critical vernacular: a continuous process of appropriation at various levels (e.g. formal, functional, significant).

From this statement, it is recognized that African architecture – in this specific case about Nigerien architecture – is the result of a rural world and a society of slow changes. It was established as the individual or family response to specific housing needs and was part of the spirit of group solutions (styles) developed for generations and based on the needs, local beliefs and materials available on the site (Preston Blier 1987; Arecchi 1998: 17–19). In such contexts, traditional constructions are the product of a "spontaneous" and original architecture in which there is no waste (Guidoni 1975) and for which everyone becomes an architect suited to the needs[6].

In the past, rural Africa was made up of small farms. Most families lived according to a subsistence economy based on agriculture or livestock (Zeleza 1993). Each small group organized its space according to the needs and hopes of life's preservation and reproduction, having a clear understanding of a world in a delicate balance between humans and natural forces. The small agricultural communities have accumulated experiences in daily relationships with the forces of nature, without practising anthropocentrism ideas about man and his techniques on the world (Arecchi 1991: 31).

The space symbolism

We argue that African architecture has been shaped and closely linked to symbolic relationships with geographical space and human and mystical culture (Fassassi 1978; Preston Blier 1987). The traditional architecture of West Africa does not exist except for a set of relationships with deities and nature (Bourdier, Minh-ha 2005: 15). The architecture was born with a spiritual and productive connection between inhabitant and habitat, connoting the existence of the group in continuity with nature (Rasmussen 1996). According to Jean-Paul Lebeuf (1967) in his African researches, architecture is a kind of script, that can be read, and associated methodologies influence how these architectural texts are ultimately understood. He says that "African housing is more than a geographical fact it is a social fact. It constitutes an extraordinary religious manifestation. It is a total phenomenon. The material, family, social, spiritual life of individuals and groups takes place there within the framework of symbolism present in every moment of existence in all parts of the house and the smallest details" (1967: 324). Correspondingly, Arecchi declares that cultural currents, social and productive structures and religious beliefs are often more important than external environmental constraints. These conditions establish the architectural forms' choices and sometimes try to overcome the constraints imposed by the use of local materials (1998: 20).

In this sense, the house is a consecrated space because it is fundamentally human and cosmic: it responds to a function, is a work of art and embodies a spiritual place (Bourdier, Minh-ha 2005).

The buildings' and villages' spaces reflect anthropomorphic precepts[7] (Fasassi 1978; Arecchi 1991: 42–56, 147–148; Guidoni 1974: 752–763), for protecting, receiving, being reborn and surviving, creating a social terrain, facilitating the communion between ancestors and divinity. Nevertheless, at the same time, the environment agrees with the energies of nature. A dwelling that is making space, a breathing element that is shaped by changing needs and therefore must be maintained or abandoned (Bourdier, Minh-ha 2005: 17–27; Preston Blier 1987). The "home" is the hub where daily life takes place and is the soul and symbol of the African family. The house, the family and the deference for older people guaranteed the maintenance of traditions. According to these norms, life was more marked towards custom than innovation (Rapoport 1969: 8–11). As Reader marks, for sub-Saharan communities "innovation and change were unacceptable risks" (1999: 263). Hence, even innovators of construction techniques were discouraged from introducing new elements (Arecchi 1991: 32).

In the sub-Saharan Africa societies, economics and power are closely linked to religion and spiritual activities. This vision is crucial in traditional African life, and also

Fig. 02.03
Village scattered in the *brousse*, Dosso region, Niger
© AT 2016

Fig. 02.04
Falwel village, Dosso region, Niger
© AT 2016

Fig. 02.05
Food market,
Niamey, Niger
© AT 2016

Fig. 02.06
Village with an evidently aggregative system of parts: barns, warehouses, meeting rooms, canopies for resting under the shadow. Niger
© AT 2016

in Niger where Islam is the official religion[8] (Bravmann 1974: 6–14; Zakari 2009).

The social organization is based on a blood relationship plan, which plays a crucial role in the choice of location and organization of the buildings within the village. Group cohesion is mainly defined by rituals which connect the worlds of the living and the dead, creating an indissoluble bond within the earth. According to African myths, man is born from clay (Lynch, Roberts 2004: 52–53) and from this material the creation of architecture flourishes, linked to the ancestors from which the protection and well-being of the family is derived (Bourdier, Minh-ha 2005).

The symbolic and religious aspects establish the structure of the settlement and the creation of a community dimension. The family is determined by its circle which brings together people, thus passing from a family (father, mother, children), to the clan, then to the ethnic group, describing the hierarchical social relationships between each member and the interpretation of the space according to rank and age (Fasassi 1978). With regard to the inhabited space, a series of "opposites" categories are identified, such as men/animals, family members/strangers, males/females, indoor/outdoor, food/lunch preparation, sleep/wake and so on (Echard 1967; Arecchi 1998: 27).

The composition of living space in the villages

Before the colonization of Niger, urban centres were located only in the Haussa region, while the rest of the country was mainly made up with scattered villages of huts.

In literature the huts are usually described only as a cone on a cylinder, and the rondavel not only dominated the pre-colonial landscape, but it also remains as popular perception of indigenous settlement (Steyn 2006: 21). This type is still found in rural areas in large parts of sub-Saharan Africa. Indeed, in Niger, Zinder and Agadez are the only two historic cities that are characterized by multi-storey houses, narrow streets, palaces, mosques and city walls since the past era.

In this context, the space composition follows specific rules: the patriarchal hierarchy of the village is divided according to the influences and statuses of the inhabitants within the community (Fasassi 1978). The habitat structure closely replicates the family organization: the genealogy, the type of marriage/relationships, the interactions between the genders and social groups. The space of living reacts promptly to any change in these components of the social composition, without the inelasticity typical of the most "consolidated" cultures (Arecchi 1998: 26). The family group conditions everything,[9] and the courtyard is the kernel of it. Schoenauer describes the "typical African round-hut compound dwelling [as] a cluster of round huts facing an enclosed central courtyard" (2000: 62).

However, the domestic architecture of tribal societies is not limited to the house but is defined as an aggregative system of parts that includes barns, warehouses, meeting rooms, canopies for resting under, market and places for prayer (Arecchi 1998: 26).

Mester de Parajd's book (1988) extensively describes how spaces are composed within both cities and villages. Equally, it is possible to read the relational dynamics through the various villages scattered around the Nigerien countryside.[10] The heart of the village is the large tree,[11] under its shadow meetings are held where only men are allowed.

The centre of the village is the square, and beside the village chief's house. Ceremonies and festivals characterizing the life of the community and the administration of justice are held in, according to the tradition where in the most egalitarian societies decisions were commonly taken in open spaces. In Niger, the elderlies sit under a canopy, covered with leafage or straw, and below simple mats or an elevated platform on average half a meter above the ground stay for protection against humidity or snakes. Moreover, where meeting rooms exist, their shape is different from the ordinary houses; in like manner the houses of the village chiefs are made more evident by palisades (Arecchi 1991: 118). Among the Haussas, the palaces of the Emirs were incorporated in to the urban fabric, creating a whole part of the city under a special status[12] (Abbado et al. 2015: 14–37).

The square is also the place of prayer[13] where faithful Muslims gather today. The religion has always permeated every activity of African society: it is considered not a way to transcend the harsh reality of life, but a "way of life" (Arecchi 1991: 109–112). The mosques are usually tiny and are located on the main square. The minaret is rare in Niger,[14] but the mihrab[15] makes it possible to distinguish the building. Sometimes, the Koranic school is placed next to it. Mosques in raw earth testify to the impact of Islam along the Savannah belt,[16] and these constructions also influence as well local structures (Abbado 2015: 58–71): the cultural context is simultaneously the product and producer of this architecture (Mester de Parajd 1988).

The market is another characteristic element of Nigerien life: it is important from a psychological, cultural and socio-economic point of view (Arecchi 1991:123), but also becomes the attraction for people

Fig. 02.07
Market in a village on the road to Dogondoutchi, Niger
© AT 2016

Fig. 02.08
Modern mosque on the road between Niamey and Gaya, Dosso region, Niger
© AT 2016

from different parts of the brousse,[17] on specific days of the week.[18]

Finally, a fundamental element of the survival of the villages is water, which is collected by women or children at the well.[19] Sometimes, that is several kilometers away from the villages, becoming the meeting point for women, who are, however, forced to walk long distances, sometimes alone.

The house is the essence of the living concept, the tangible manifestation of a complex value system that combines cultures, alliances, climate, resources and settlement geography. Generalizing the traditional African architecture, the "rondavel" is a model of rural architecture widespread in the whole of Africa: several huts – each is a room – are built for a particular function around a central courtyard, the lapa or lolwapa or lelapa (Oliver 2003: 158).

As Walton writes the *lelapa*, enveloped by a screen of reeds, was actually the real home, the hut itself served only as a bedroom and storeroom (1956: 52). Rapoport emphasizes that circular huts are easier to cover than rectangular ones, but he stresses that the choice may ultimately depend on the symbolic nature of the forms, in fact some traditional cultures do not have a word for "straight" (1969: 25,77).

However, as pointed out by Steyn (2006) the rondavel offers an almost unlimited flexibility to respond to the evolving changes of the clan.[20] The root of the village is composed of enclosed courtyards and multiple housing, which usually correspond to the clan's components:[21] social relationships materialize through the aggregation of individual homes. The inhabited space becomes an accurate description of society, and the idea of "home" is the image of a clearing or a courtyard (Arecchi 1991; Arecchi 1998; Bourdier, Min-ha 2005: 37–41). Most of the activities take place outdoors and the interior spaces assume meaning in relation to the open space (Bourdier, Min-ha 2005: 28–37).

The module is composed of the fence[22] that encircles the shade of a tree. Also, in the courtyard, there is a specific area used for cooking, which includes three stones on which a large muffler is placed.[23] Cooking, dining and other activities usually take place outdoors, due to the high temperatures reached during the day.

Therefore, shade and natural ventilation become essential for the activities. For this reason, canopies are built to protect from the sun, under which women clean the millet or prepare food. These sheds, placed next to the house, can be covered with mats or woven wood, and the fodder is stored in them. Bed structures of wooden bundles are laid below, they are also

Fig. 02.09
Well in the country-side, Tillabéri region, Niger
© AT 2016

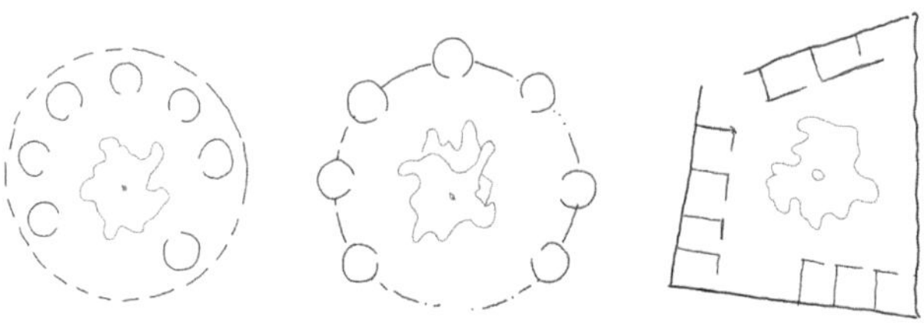

Fig. 02.10
Different composition of living space in the villages © AT 2020

Fig. 02.11
Village near Salt Lake, Dallol Foga, Dosso region, Niger
© AT 2016

used during the night to sleep outside especially during the summer season. Alternatively, mats are positioned as rugs to create a pavement.

Inside the courtyard, warehouses and befitting fences for animals, which cohabit in the same space, are sited. In one of the corners are the toilets, which are usually in the brousse and consist of a hole in the ground with a woven straw mat rolled around.

A fundamental part of the villages are the barns, which are recognized from the houses by their typological characteristics, different from region to region. Some of these are scattered in the fields, while others are located on the perimeter or in the centre of the villages.

Lastly, the African house is in harmony with the order of the universe: it is built in compliance with the ritual rules and constitutes both a real and symbolic space, with signs of the sacred and tradition. Its decorations are not purely aesthetic but are a "writing" which create the house as a book, an archive of traditions and history (Arecchi, Delisse 1999).

Types and constructive characters

Despite the variety of cultures within this region, there are common attributes that are identified across multiple territories and suggest a better understanding of the environmental character built in West Africa. Traditional Nigerien architecture has been formed over millennia, settling the parameters which interfere with its creation. As in nature, traditional architecture has been shaped through a selection process, through slow transformations of the species. It has adapted to the physical and cultural changes of space and inhabitants.

However, western colonization led to an upheaval in these traditional societies, and subsequently the phenomenon got even worse due to the decolonization, with the loss of the identity and traditions. The developing countries are now leaning towards progress and modernization, which they interpret as an opposition to return to the origins. In the architectural field, the outcomes are the loss of building traditions and builders' ability to work with traditional techniques and materials.

The parameters, which determine traditional architecture, are objective – related to the environment – and subjective – related to culture. The interpolation of these factors determines different ranges of complexity.

The "objective" parameters are related to the geographical and climatic environment, which must be taken into consideration. In particular, indigenous societies enhance their ability to respond effectively to climate change when they draw upon their traditional ecological knowledge (Daoudi et al. 2019: 119–154)

Secondly, the building materials were based on the available resourses which nature had made readily available site by site. Existing geographical conditions played a vital role in the availability of these building materials (Bello 2001). These naturally occurring materials include clay, mud, timber, bamboo, thatches, grasses, palm frond, raffia palm, leaves, bark of trees, local fibre of bush palm, animal waste (dung) (Odeyale, Adekunle 2008).

Know-how is a fundamental factor for determining all the available techniques. It depends on the degrees of cultural development, the available tools, inventions, creativity and external influences.

Lastly, the lifestyle determines substantial choices, such as for nomadic, semi-nomadic and sedentary populations. Neighbourhood and security relationships also define plan, location and structure preferences (James-Chakraborty 2014).

Instead, the "subjective" parameters are linked to the cultural aspect and define the aesthetic values, following the cultural principles of the group. Analogous to nature (Arecchi, Delisse 1999), traditions are formed traditions, which guarantee the balance in social groups and determine the magical and religious principles. Finally, the magical environment influences the habitat to make it conform to the rules of life and protect it from the supernatural and evil spirits.

Traditional dwellings: classification

Analysing the typology of dwelling in Niger, it is evident the different ethnic groups[24] have defined the peculiar character of living space. However, similarities can be seen between the different types, due to the interactions that have taken place over the centuries. The history of the Nigerien peoples cannot be separated from those of the neighbouring territories, including Mali, Nigeria, Burkina Faso, Togo and Benin.[25] Consequently, the different districts of Niger also have similar building types.

Nomad houses

Tends and mobile houses | vegetable materials and leathers

The presence of the Sahara Desert, which occupies most of the Nigerien territory, has also dictated the organization of human life. Groups of people became the connecting elements between the two North-South banks of the great "void". Therefore, the desert gave a rhythm to the nomads' life, where the greatest needs for sustenance were water, grass and salt (Bidi 2000: 93–96). The seasonal displacement

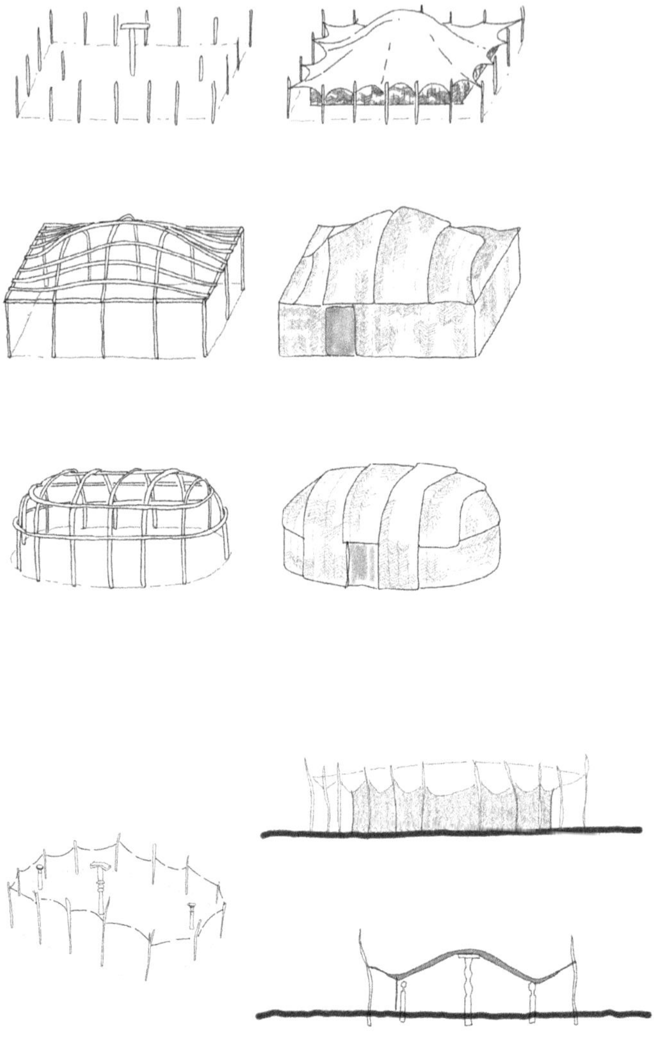

Fig. 02.12
Nomad houses: tends and mobile houses with vegetable materials and leathers
Source: graphic eleboration by the author based on Mester de Parajd 1988.

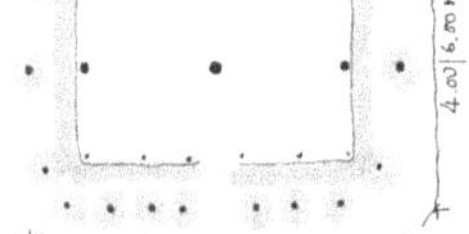

of transhumant nomads follows established paths, influenced by climatic conditions and the social behaviour of each population.

The transhumance's distances and the timing for the displacements create a different community identity. The higher distance produces a more significant number of people who undertake migration. For these reasons, various nomadic societies, who migrate frequently, have invented easily detachable and transportable houses to relocate according to the climate and new conditions (Prussin, Adan 1995). Moreover, nomads who carry out extensive displacement once a year have permanent homes in either of the seasonal camps (Arecchi 1991: 42). This situation gives rise to different types of nomadic dwelling: from the tent – as for the Tuaregs or groups of Berber nomads – to the hut for the unity family, where according to the typical patriarchal structure the head of the household lives with three or four wives, and their married children (Arecchi, Delisse 1999: 40–42).

In desert areas, Tuareg nomads[26] live in tents. The Tuareg tent is crucial in ritual and symbol, as well as in subsistence, property relations and marriage: the tent is the centrepiece of the Tuaregs' culture in both nomadic and semi-nomadic communities[27] (Rasmussen 1996). It is aggregated in camps, which in extended stays are equipped with enclosures for animals (Guidoni 1975: 347). Usually, the plan is rectangular; the frame is formed up of 2/4 parallel forks which support horizontal poles; sometimes instead of the poles, there are arched wooden structures. The cover is made of leather or stretched fabric (Guidoni 1975).

The leather tent is the typical habitat of the Tuareg nomads, located in the west of the country and of the Iwiliminden tribe, located to the west of Agadez. Almost a hundred goatskins and sheepskins are sewn together with leather threads and waterproofed with animal fat.[28] Rebars, consisting of vertical uprights and a central pillar often carved and pyrographed at the top, support the leathers. The tent encloses an internal open space which is covered with vegetable fibres and leather rugs, useful for protection from the wind, sandstorms and also for privacy. If the leather is undecorated, the peripheral carpet incorporates refined geometric decorative motifs. The interior is occupied by a movable bed, rugs, leather cushions and vessels on supports (Mester de Parajd 1988: 22; Taton 1984). Typically, one tent is used by members of the same family. As noted, the tent is mainly associated with women, who build it in about four hours (Taton 1984), and it is their own. The motherly tent is, in fact, at the centre of the rites of passage:

Fig. 02.13
Stable huts with vegetable materials, Gaya, Niger
© AT 2016

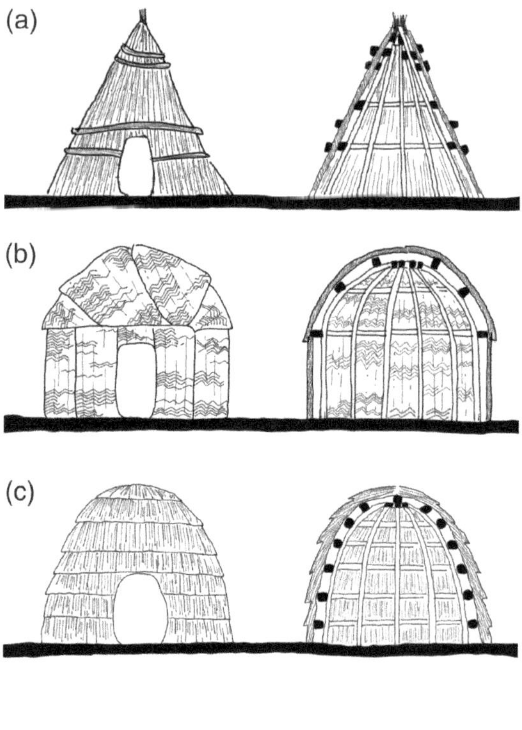

Fig. 02.14
Typology of huts with vegetable materials
(a) conical hut
(b) hut in woven straw mat
(c) straw hut
Source: graphic eleboration by the author based on Mester de Parajd 1988.

weddings, children's name days and mortuary rituals (Rasmussen 1996).

The rectangular tent in sewn rugs is the kind of tent by the Tuaregs on the East of the country and the Toubous, but there can also be examples in circular or elliptical form.[29] It is built with a wooden reinforcement comprising vertical poles and horizontal bars that create the walls on four sides. Three or five vaulted arches are planted in the ground and link the upper sides supported by a secondary wooden frame. Sewn vegetable fibre rugs cover the whole (Mester de Parajd 1988: 22). Another typology is the Iwarwaran rectangular tent of the sedentary tribes of Kel-Gress and Kel-Aïr (Tuareg), who live mainly in the regions of Ader and Aïr. The structure is made up of four vertical struts mounted at the corners and inserted into the ground. A principal crossbeam at the top and others in the middle join the pegs together. Sometimes two or three central trunks can be placed along the centre of the room. Transverse arches rest on the central crossbar, where branches are placed perpendicularly to shape the roof. The obtained shape recalls a dome. Palm mats (2/3 m x 10/12 m on each side) are laid on the roof and the walls, fixed to the structure by ropes. The use of mats guarantees proper ventilation of the tent (Taton 1984).

Generally, this type of housing is designed for nomadic people who consider the roof only a refuge for the night, since they carry out all their daily activities outdoors. Family shelters consist of aggregations of tents surrounding an open space for meals and community life (Arecchi 1991; Arecchi 1998).

Stable housing

In the Sahel belt, two typologies of the hut can be distinguished which are widespread throughout the territory of Niger, except in desert areas: the hut wholly built with straw[30] and the hut in raw earth or banco. Usually, the most straightforward method for covering a circular room is a conical thatched roof, borne by a central pole. Depending on the region or the availability of vegetation, materials, traditions and climatic requirements, it is possible to find different variants, entirely or partially made up of vegetation or earth. At the same time, the use of stone is rare, save for a few exceptions.[31] The single huts also can set a "beehive" aggregation,[32] considered the oldest African style. Alternatively, they may be disconnected and circular huts, with a diameter equal to or greater than the height.

 # Stable hut | vegetable materials |

The huts made with vegetal materials mostly use mats, straw, wood and leaves.

The hut with dry mats is prevalent in most of the country: it is generally rectangular and has an internal wooden structure. It is a temporary house, usually built after the harvest and in between the rainy seasons. The mats let the rain pass through, and for this reason, before the new wet season, the hut is made completely made watertight (Mester de Parajd 1988: 25–26; Arecchi 1991; Arecchi 1998).

The classic straw hut is covered with woven straw or millet reeds. It is widespread, especially in the villages in the Djerma-Songhaï area, and ensures excellent thermal comfort, as well as being economical, even if it only lasts four years.

A variant of this is found in the Gourmantchés tribe, south of Niger and also in the Haussa regions: a wooden structure on the external perimeter has the benefit of protecting the facade (Mester de Parajd 1988: 26; Arecchi 1991; Arecchi 1998).

The Zarma and the Peul in the centre of Niger used to live in huts composed of a central trunk where different arches and transverse struts are fixed. The walls and roof – with a tip on the top (about 0.5 m high) – are covered with woven straw and palm ropes or millet stalks (Taton 1984; Bourdier, Minh-ha 2005: 173–183).

Huts are assembled with palm leaves instead of straw where plantations of date palms are more common, as in the eastern areas of Zinder (Mester de Parajd 1988: 26).

 # Mixed technology housing | earth and wood |

The Haussa population lives in areas characterized by a tropical climate with a rainy season from June to September and a dry season for the rest of the year. According to the environment, vernacular architecture has developed using earth combined with straw, rice hulls and other vegetable materials as aggregates. Usually, the roofs are built with dried straw, and palm trees constitute structural wood. The foundations rest directly on laterite (Houben, Guillaud 2006: 211), which constitutes the predominant type of soil in these regions. The dwellings can be circular or rectangular, with conical or flat roofs. Attention to the climatic aspect and ventilation leads to the provision of small windows, which help regulate heat gain (Mester de Parajd 1988; Arecchi 1991; Arecchi 1998).

#Mixed house of straw,[33] banco and wood frame

This type of house is widespread in Niger, especially in the Haussa country. Traditionally, the banco is hand-modelled based on wooden reinforcement, but today the wall is usually created with sun-dried bricks.

There are several variations of wooden reinforcement use. It can be

Fig. 02.15
Houses of straw and earth, Tillibéri region, Niger
©AT 2016

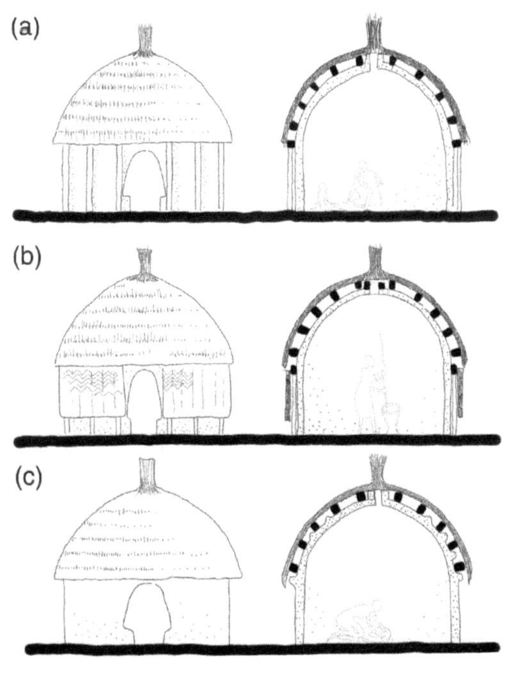

Fig. 02.16
Typology of huts with vegetable materials
(a) clay hut with wooden frame
(b) clay hut with wooden frame and straw nat
(c) clay hut
Source: graphic eleboration by the author based on Mester de Parajd 1988.

Fig. 02.17
Village near salt lake, Dosso region, Niger
© AT 2016

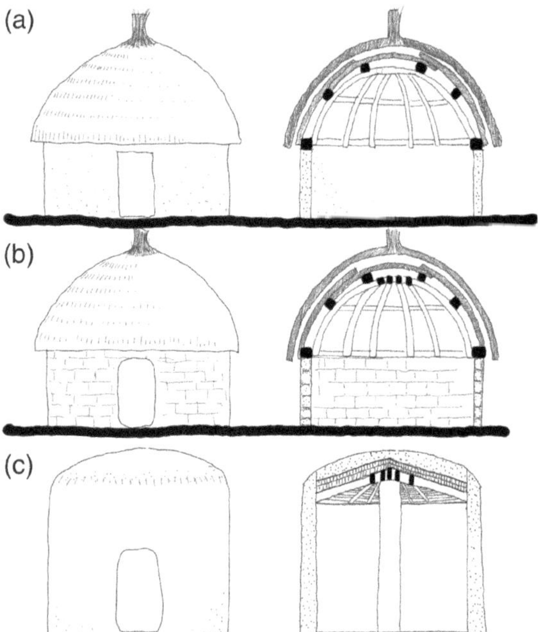

Fig. 02.18
Typology of huts with vegetable materials
(a) clay hut with internal wooden frame
(b) hut with earth blocks
(c) clay house
Source: graphic eleboration by the author based on Mester de Parajd 1988.

visible inside the house or outside, ensuring better protection for the banco during the rains, thanks to the wider roof. The external wooden reinforcement can be covered with straw mats, which protect the banco wall from the rain and the air gap of a few centimetres in between provides excellent thermal insulation.

Sometimes, in those regions where stones are abundant, a variable proportion of these can be put in the banco wall (Mester de Parajd 1988: 26–27; Taton 1984).

#Raw earth housing

The banco house is the most common in Niger as in the neighbouring countries, especially in regions inhabited by sedentary populations. Traditionally these dwellings are made with a circular plan and thatched roof (today only in the villages in the countryside), or quadrilateral with a flat roof (in the cities or the urban centres). By observing this style, different variations of the typological principle can be highlighted.

In a detailed analysis, we highlight circular banco huts with a thatched roof: the outer cover protects the facade from the rain and the earth provides excellent thermal insulation. The Haussa rondavel in the Tahoua region is improved with a central pillar also of banco and in several regions variants depend on the external wooden reinforcement or on the straw surfacing which allows better protection from the rain and increases the insulation.

The second solution is the banco house with a terrace roof. Originally from the Haussa region, this typology has spread since the beginning of the century throughout the territory of Niger,[34] mixing a Haussa technology with various local construction techniques in the process.

This typology has largely been used thanks to the square or rectangular plan, which permits an easy internal organization of space. Besides, it gives a high thermal comfort in the various periods of the year. During the dry season, the banco is a thermal flywheel and replaces the freshness gained during the day, indispensable for high temperatures in the dry season.[35] Furthermore, the durability[36] of the habitat is warranted if care is taken in stabilizing the raw earth and constantly in preserving it.

There is a disadvantage related to the flat roof, which needs long and straight wooden beams[37] that are difficult to find in the Sahelian area. The flat roof risks being broken through if not well maintained over time, especially in the southern regions, characterized by more abundant rainfall. To solve the flat roof problems in the Tahoua district, sometimes a banco pillar is added in the middle of the house to reduce the load-bearing capacity of the beams. Then a rounded roof has been used, which improves the drain of the rain.

The defining characteristic of the Haussa vernacular tradition is the ribbed vault, also called false Haussa vault,[38] which consists of reducing the long beams and replacing them with arches composed of bundles of wood connected and plastered with banco (Arecchi, Delisse 1999: 55). Mester de Parajd called this technique the "Sahel armed banco", which allows the vaulted roof to be divided into reduced bars. Pieces of a wooden junction are placed between these ribs and mats are laid on them. Finally, the structure is covered with banco (1988: 46–53). The Gugaa (Haussa vault)[39] has a ceiling made of coffered branches, supported by eight stones, which reinforced the base (Arecchi, Delisse 1999: 106).

The flowering of decorative art on the plastered facades of the banco houses is worthy of attention. The decoration is a vector of the society's value, recovers the identity of the community and communicates symbols of the group's morals or ethics. The earthen architecture is richly decorated and it takes on an aesthetic that is magical, religious, apotropaic but also has utilitarian value. Observing the decorations and the renewal of the decorative style usually reveals the relationships and influences between peoples and their cultures (Houben, Guillaud 2006: 347). In African architecture, the decorated components are the entrances of the houses, the barns, the canopies, the sacred and community places, the women's rooms, the doors, the internal walls of the rooms, the pinnacles of the roofs (Arecchi, Delisse 1999: 56; Bourdier, Minh-ha 2005: 26). In these places, the decoration expresses various styles: shapes and colours with a variety of natural pigments, reliefs on the surface which create shadows and light. In particular, we want to underline the floral or geometric reliefs[40] in the urban dwellings of Niger or the Haussa and Dogon villages, and the painted geometric decorations, modelled or carved in the thickness of the wall. Furthermore, the geometric relief with formed shadows improves the thermal conditions and reduces the leaching, increasing the life span of the wall (Mester de Parajd 1988; Houben, Guillaud 2006: 347).

The plasters are marked by different techniques with different textures, or granulometry, such as smoothed finishing layer or reliefs sculptured on the wall or in the thickness of the plaster, which can sometimes be coloured.[41] The external clay-based plaster, sometimes mixed with néré[42] oil, guarantees better protection from the rain (Mester de Parajd 1988: 45) According to the Haussa tradition, the plasters should be combined in seven

Fig. 02.19
Ribbed vault, also called false Haussa vault. Palais du Djermakoye in Dosso
© LS 2017

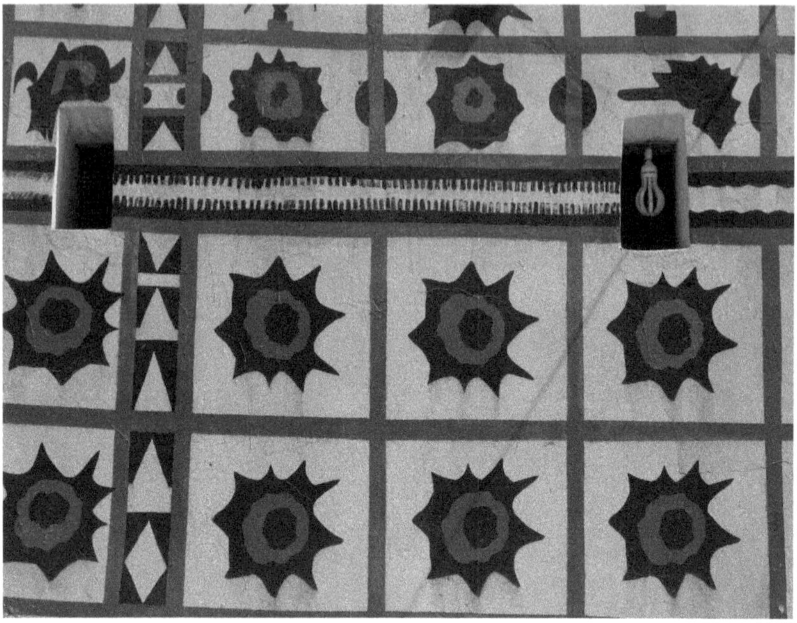

Fig. 02.20
Palais du Djermakoye in Dosso
© LS 2017

different ways: earth and straw with potash, used for making tinctures, or a blend of local locust beans (Arecchi, Delisse 1999: 55).

In general, the external facades are left slightly rough as they are less sensitive to cracking or defects on the substrate, while the internal surfaces are often smoothed to reduce the amount of dust that settles due to the daily life activities. The different treatments of the surface are carried out both before the plaster hardens and after.

The decorations are painted on a flat facade and can cover a large part of the facade. The artistic motifs are traditionally geometric patterns or stylized flowers, inserted on drawn squares, or sometimes animal figures.[43] In Zinder, elaborate geometric patterns and woven patterns are traced in the thickness of the walls (Arecchi, Delisse 1999: 55). In Agadez, the motifs can also highlight the verticality underlining the structural elements. Indeed, this vertical decor is possible where the rains are less intense, while the woven decor is more often used where the rainy season is heavier.

Barns

Barns are a constant element in the Nigerien landscape, scattered throughout the brousse. They are mostly used to store millet and sorghum, and have the same technical peculiarities as the dwellings, but they are always slightly raised from the ground, which makes them instantly recognizable in the landscape. Stones or wooden poles are placed on the ground to support a platform formed of large branches or flat stones. This platform holds up the granary which can be either raw earth or banco (Carrol 1992: 19–27).

The barns are the symbol of each of the ethnic groups and their constructive traditions and are expressed in a multitude of different shapes and variations. Mester de Parajd (1988: 53–60) compiled an exhaustive list of the typologies located in the Niger country.

The most common granary (Haussa) is made of smooth mud and a round shape, which can reach up to 4.5 m in height, but the thickness does not exceed 10 cm. It is laid on a stone base, which raises it from the ground and has a hole in the top (reached by a staircase) and sometimes on one side. Haussa granaries are always made of mud in a cylindrical shape with buttresses which allow a thin wall thickness.

The cone-shaped granaries with a wooden structure platform supporting a covering of branches and millet canes are scattered in the Niamey district. Alternatively, the hemispherical shape (Djerma region) with a wooden structure has bands of wooden rods on which mats are laid. Alternatively, the cylinder-shaped granary has a structure of wooden

Fig. 02.21
A type of barn, Gaya, Niger
© LS 2017

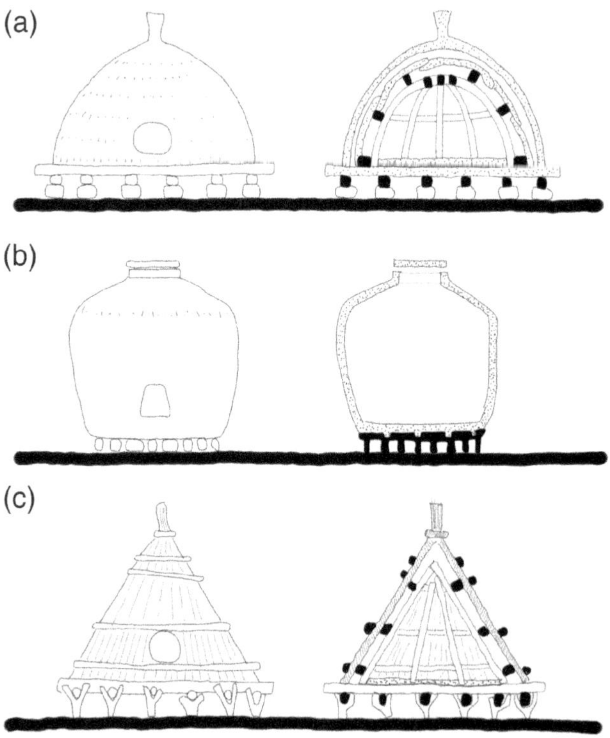

Fig. 02.22
Different typology of barns
(a) Hemisphere clay barn
(b) Clay barn
(c) Conic barn with wood and straw
Source: graphic eleboration by the author based on Mester de Parajd 1988.

arches and is covered with bundles of woven millet (Haussa, Zinder countries).

Along the River Niger, towards Tillaberi, the mud barns are circular, smooth, decorated with annular rings along the facade and rarely exceeding 2.5 m–3 m in height. They have a hole in the top and one on the side for collecting storage materials.

Towards the border with Mali, there are round-shaped barns, bristling with stones, which are used to climb up to the top hole. The dimensions are approximately 2.5–3 m.

Lastly, in the South, where rainfall is abundant, the warehouses are covered with bundles to give them protection from rain and heat. Inside the barns, some clay jars, typical of the areas south of Dosso, can be found.

Openings

"... what is needed at the beginning of the new millennium is an architectural perspective in which valuable vernacular knowledge is integrated with equally valuable modern knowledge" (Asquith, Vellinga 2006:18).

This study does not propose ideological solutions where vernacular architecture is suggested as the only possible outcome for buildings in emergency contexts. Conversely, we wish to understand how the principles of traditional architecture have the correct answers to living in challenging environmental contexts. Over the years, the overlapping of knowledge and answers has therefore defined key opinions that should become the basis for setting up new structures too.

According to Steyn (2006) and Odeyale and Adekunle (2008), the "growth" or "development" of traditional societies, the stated objectives of transformations which are taking place, evolve in reality into their destruction.

Currently, we are faced with double cognitive poverty. The first concerns the loss of "vernacular" skills, which defined the characteristics of architecture based on climate and resources: raw earth and a thatched roof were solutions of necessity but also efficient answers. Instead today, the ancient methods are commonly rejected by the population, which considers those techniques to be an expression of poverty and backwardness. The second poverty concerns the intention to be modern using western imported materials. The use of these materials is inappropriate and the metal sheets and concrete bricks that are so standard now in Niger don't solve the problem. The metal sheets used for the roofs amplify the heat, preventing people from living inside their homes. Cement is preferred for its strength and durability to raw earth. It is utilised in hollow blocks, which guarantee economic, easy and fast realizability, but have technical properties unsuitable for the desert climate of the region.

Although the traditional building techniques survive in the most remote villages scattered in the countryside and along the Niger's main routes, today's tendency to build "as the Westerners" leads to abandoning the traditional spatiality and techniques. The buildings heat up very quickly, due to the harsh climatic conditions, especially during the hot season (March/June), preventing sleeping inside the houses at night, also because few windows make it difficult for the air to circulate.

We cannot expect to "stop history" by following the utopia of integral and museological conservation of disappearing cultures or their existing environments. We want to propose a classification of "endogenous" cultural development lines, which do not lead to the separation between profound cultural heritage and new habits of life, acquired in the era of industrialization. The aim is to start with the ancient assimilated knowledge and unlock ideas, explorations and identify new approaches to sustainable and innovative techniques, which could be interpreted and used by those with local knowledge, practice and wisdom, as well as by recycling waste materials put back into new cycles (McDonough 1996; Braungart, McDonough 2002).

Today's studies on vernacular habitats show an interest in more than their static "image". Unlike previously, we opt to highlight critical, creative and procedural aspects of vernacular examples. This certainly transfers the vernacular to a prominent position in architectural research, as well as in practice and education. Local solutions have been refined over the past years by culture and social logic (Rapoport 2005; Oliver 2003), thus achieving application in an appropriate context. Many of the criteria defining modern design are inspired by traditional or vernacular solutions in which social, cultural, spatial, physical, technological and aesthetic factors are combined in a single complex definition (Glancey 2003). "Shape is not the aim of the work, but only the result" (Fromonot 2003: 23) and this outcome is based on some typical dynamics. Components such as material sensitivity, suitable and technological climate solutions, tactility, minimal systems, modularity, environment adaptability, as well as temporality or fluidity, define a boundary between traditions and variable modernity.

Mapping the similarities in

principles and recognizable spatialities (Eglash 1999) makes it possible to transmit concepts and techniques from traditional to modern or from modern to traditional in a bi-directional process. The objective is not to enclose Africa in its traditions, but to seek new ways to achieve the aspirations of the inhabitants of Niger through the promotion of cheaper and more efficient local materials. Modelling the traditional Sahelian architecture with a fusion of innovation will allow us to face and answer difficult questions, and to guarantee a necessary architecture for everyone, starting from the awareness of the past techniques and transforming traditional materials into a recycled modern use.

>>> Notes

1 As Prussin reports, the evolution of thought concerning African architecture is fascinating. Just a few years ago, an American magazine sent a troupe of photographers to Africa to document the great eras of African history, taking on monumental architectural illustrations. On their return, the comment was: "All we could find were a bunch of mud huts!", thinking that vernacular architecture is lacking in both identity and meaning. Their reaction was the echo of the same feeling generated over a century ago. It is relevant to note that three eminent explorer-travellers, each of whom reported their experiences and impressions from different parts of West Africa, had both used an identical design to illustrate indigenous homes. The plate appeared for the first time in Eugene Mage *Voyage dans le Soudan Occidental* (Paris, 1868). The same appeared a second time in J. A. Skertchly *Dahomey As It Is* (London, 1874), to illustrate a Mahi village in the north of Dahomey, and a third time in Lieutenant Pietri *Les Franfais au Niger* (Paris, 1885). Therefore, although the reality in the three regions was very different, the reportage reflected the European mental image of the times: everything was identical (Prussin 1974: 143).

2 Some historians and conservationists of traditional architecture use "vernacular" in a condescending way to refer to poorly conceived, personalized or watered-down versions of high style. This work is often called "vernacular expression or manifestation" of a high-style work or "my personal" "vernacularized". Besides, architects currently use the expression "neo-vernacular" to refer to a wide range of contemporary projects that elaborate contextual or regional elements or use popular forms as a starting point for

modern design (Heath 2003).

3 His first significant publication on architecture, *Shelter and Society*, appeared in 1969, when he was working at the Architectural Association. The main goal of the book was to indicate how all those traditions were built to "meet the needs of their respective communities and contain values special to them" (Oliver 1969: 31). Oliver further explored the intricate relationship between the architectural traditions and society they formed part of during the 1970s in his subsequent publications. *Shelter in Africa* (1971), *Shelter, Sign & Symbol* (1975) and the little-known *Shelter in Greece* (1974, co-edited with Orestis Doumanis).

4 http://www.developmentofpeoples.org/projects/case/547/une-industrie-culturelle-en-developpement-larchitecture-en-terre.

5 "An African man may describe a circle on the ground with a stick at arm's length, which will constitute the plan of his dwelling, and he may mark out the boundary of his yard. However, when he builds, he will use traditional forms, employ the technology of his ancestors and may have his hut plastered and decorated by his wife, with designs that are symbolic to his people and immediately identifiable with them. A limited degree of design is involved, but the architectural solutions have been arrived at over generations" (Oliver 2006: 4).

6 According to Olaudah Equiano, known as Gustavus Vassa, Nigerian writer and activist during the eighteenth century and naturalized English, the Igbo's folkways were to collaborate with all the village's neighbours for the construction of a new house and a great party was the reward (Curtin 1967). This culture could refer to several other traditional groups. Moreover, the majority of the rural societies in Africa decorate the exterior of their building with painted designs or with relief (raised) patterns worked into a soft clay surface. The decoration of houses is mainly a women's responsibility (Arecchi 1998: 25; Odeyale, Adekunle 2008; James-Chakraborty 2014: 412).

7 For example, Arecchi studied the plan of Dogon villages, which should represent the position of a lying man with his head turned to the North. He reports that a large square represents the head, the feet are the altars of the various families; the hands are some circular huts in which women retire during menstruation; the family homes are the chest. This anthropomorphic design is underlined by a conical altar and a millstone with a hole, which represent the male and female sexual organs (Arecchi 1998: 46).

8 The 2001 national Census determined that 98.7 per cent of Nigeriens are Muslims (95 per cent Sunni and 5 per cent Shiite). The remainder are Christians, animists and other tribal creeds. This situation means that Niger has the most Islamized population in West Africa after Mauritania. The first contact with Islam occurred only 15 years after the establishment of the religion, during the subjugate of local chieftains by the Muslim generals. The conquest took as a result of influxes of Berber warriors who were able to react to the problems linked with the integration of different cultures. However, various animist cults remained for a long time alongside the Islamic religion, contributing to the creation of hybrid rituals. In an apparent paradox, it is the French occupation of Niger that ended up giving a decisive boost to Islam in this part of Western Africa (Bravmann 1974; Idrissa 2020: 282–286).

9 Fassassi (1978) shows how each region, city or house is different, every life has an

influence, all the details of African urbanization are regulated by a small variable from which researchers can draw common characteristics. Humans strongly influence their environment.

10 The layout of the villages in the settlements and the typological architectural characteristics were the object of the research survey in Niger carried out by the author between February and June 2016.

11 The tree is the sacred emblem of the local divine principle (*genius loci*). Traditionally, the tree was always present at the founding act of the village and remains a sign of divine protection of the village chief and the village itself. It has become the symbol of the foundation, the landmark, the guest who welcomes the group into new territory (Guidoni 1975: 302).

12 Arecchi recognized that the rondavels are not a recent construction among the Haussa. An influence subsists due to the emirs of the Fulani tribe, who since the Holy War have taken possession of Niger's territories (1991: 122). Even today, in some areas of Niger (e.g. Dosso region or Agadez district) there are villages where houses are traditionally square and built on banco. Denyer points out that in some parts of Africa, rural houses were traditional thatched huts, while urban dwellings were cubic (1978: 165). The rectangular plans are, as exposed by Hillier and Hanson (1984: 163–175), logically more suited to dense aggregation ("connecting units") with courtyards defined by cell walls. Besides, many rural populations undeniably admired rectangular western forms, which they considered a symbol of progression (Steyn 2006).

13 For Muslims, the five daily prayers can be recited in different places, even during travels. Therefore, the religious sites can be anywhere, whether in the areas bounded by stone or banco walls, or in mosques built as simple huts.

14 The Agadez minaret is the most impressive in the world built entirely of terracotta: it rises to 27 m in height, with a truncated pyramid (9x9 m at the base and 2x2 m at the top) and is reinforced by horizontal palm wood beams. These wooden beams extended 1–1.5 m from the façade surface and are useful for the maintenance of the tower. The mosque, built in the sixteenth century, is also wholly constructed on earth (Abbado et al. 2015: 60–61). Another characteristic example was the ancient mosque of Dosso (1917), now demolished. The Djermakoye (King of Djorma tribe) employed Haussa builders for its realization, who built it on raw earth and in the thickness of its walls the relics of the Marabout were placed (Mester de Parajd 1988).

15 Recess installed in the middle of the back wall in a mosque designating the direction of Mecca and where the Imam stands during Friday prayer (Larousse 2021).

16 Although the Niger's population is 98 per cent Muslim, except for some animist or Christian groups, curious is the coexisting survival of ancestral beliefs in different forms: as is the sorcery during diseases, for good luck or against bad luck or witchcraft. Even the fetishes objects are traditionally utilized during the construction of dwellings or various ceremonies. Amulets are considered elements which keep the spirits away, and for this reason are constantly present in each house.

17 "Brousse" is the typical word which the locals use for defining the countryside.

18 The markets have a very different character depending on where they are located: generally, in the countryside they are

under a tree at the crossroads; in the urban centres along main routes they are located on the edges of the streets. On the outskirts of villages there may be empty plots used as a market. In particular, the animal market is outside the residential area, and the traders can camp close to it during market days. Unlike in urban areas the market is within the city and occupies an entire neighborhood (like the Market of Niamey and Dosso).

19 The groundwater level in Niger is very deep (about 100 m), for this reason, the construction of the well constitutes a large financial commitment. In the villages, in particular, the water is still being collected manually by women via a system of winches.

20 The rondavel responds to different social structures (monogamous families are more compact than polygamous ones), economic activities (spatial relationships with crops, granaries, kraal for cattle and cattle pens) and external threats such as predators and enemies (protective fences). Not only farms but also villages and groups of villages can be contracted or expanded to form molecular or carpet typologies (Steyn 2006)

21 The layout of fences in settlements varies according to how they are built and the ethnicity. For example, the Peul's huts are located inside the hedge. In the same area, the Gourmantchés arrange the huts so they straddle the border. In high-density urban areas, the house, usually a rectangular square, is attached to the fence around it (Mester de Parajd 1988).

22 Different materials can be used to guarantee the privacy of the family. Boundaries are made with wooden poles and mats, where millet branches are intertwined between the wooden pegs. If the huts are mostly built on banco even fences are made with the same material, usually realized by hand or with dried clay bricks, sometimes enriched with stones. This technique is adopted in the urban centres more than in villages, but in cities like Niamey, the raw earth is today commonly replaced by concrete bricks.

23 Food is still cooked through wood combustion, as the use of gas cylinders is culturally refused. The harvesting firewood for cooking, especially in the villages, is one of the main factors that are affecting deforestation. Besides, collecting wood is the responsibility of women, who spend a long time away from the group with solo walks for hours in the bush, increasing the Sexual & Gender-Based Violence (GBV) risks as reported by the Sustainable Development Goals (SDGs) Agenda 2030.

24 Summarizing, sedentary ethnic groups represent four-fifths of the Nigerien population and usually live in the regions in which they originated. Haussa is the most populous group which lives in the centre of Niger; Songhaï and Djerma in the West; Kanouri in the far East and Gourmantché in the far West. Nomads mostly live towards the northern borders. They are Touareg (Kel-Aïr towards Agadez and Iwiliminden on the West of Tahoua), Peuls are scattered throughout the territory (not only in Niger), Bororos in the region of Dokoro and Mainé-Soroa, Arab groups and Toubous in the East (Taton 1984; Gallais 1984).

25 Originally the Songhaï Empire extended from Gao to Timbuktu, to present-day Niger. Another group of warriors from the Djerma-Zarma society lived in the southern regions of Tillabéri and Dosso. Later, these two ethnic groups merged, giving rise to the Djerma-Songhaï, which is the second

most important group in today's Niger (Arecchi 1999: 37). The Haussa city-states were founded before the year 1000 in the North region, on the confluence of the Benué River and the Niger. At the beginning of the fifteenth century, with the spread of camels, these cities became the major traffic centres across the Sahara. With the arrival of the Europeans, the slave trade changed the economic and strategic assets, and also the Haussa cities (e.g. Kano, Katsina, Zaria) became the commercial centres along these routes. The same language and Muslim religion are common in the Haussa towns: the mosques are the core of the cities, along with the Emir's palace and the market, while districts inhabited by specific ethnic groups are all around (Arecchi 1999: 42).

26 The Tuareg are well known as the "blue men of the desert", for their indigo-coloured face, due to their turban. This ethnic group has uncertain origins and is part of a community of Berber-speaking dialects that never formed a state. The Tuareg form a fragment in a cultural and linguistic universe. Anthropologists now admit that the Berbers descended from proto-Mediterranean groups from the East in the eighth millennium and slowly spread to the Maghreb and Sahara. Subsequently, they were largely Arabized. It was the Arabs who gave them the name "Tuareg", although different interpretations concern the roots of that word. It is usual to identify a Tuareg as one who speaks Tamasheq and adopts the lifestyle of this ethnic group: one can become Tuareg or be born Tuareg. The singularity of this definition foreshadows the difficulty of identifying this population. Since African independence in the early 1960s, the Tuareg have been divided into five states. They have become citizens of multi-ethnic nations but mostly live in the Sahel States. In Niger and Mali, the Tuareg represent a significant percentage of the population, respectively 10 and 15 per cent. However, the Tuareg do not currently form a majority cultural community in any state and are always moving, even though over time they have absorbed part of the cultural traditions of the countries they have joined (Ramir 1991).

27 The nomads prefer the use of tents because of their belief that they would get sick if they should spend too long in a fixed home. However, some Tuareg groups are sedentary farmers, who live in square houses of stone and mud, beehive, or even semi-underground. Also, in the Tuareg populations the marriage ceremony is called "éhen", which means "tent" (Arecchi 1991: 40).

28 Goat or sheepskins are treated with butter and red ochre, for waterproofing and protecting it from sunlight. This treatment gives the traditional intense red colour. The constant maintenance involves replacing the pieces that are in a bad state of conservation and allows the tent to have a life span of about ten years. (Arecchi 1991: 40; Taton 1984).

29 The Wogos tribe, along the Tillaberi River on Mali's border, build tents with stitched rugs that are decorated with large light and dark bands (Mester de Parajd 1988: 22).

30 The straw hut is the oldest typology and is characteristic of the Zarma or Peul peoples (Taton 1984).

31 The use of stone is rare in Niger due to its limited presence in the territory, composed mainly of laterite. The only province where examples of stone buildings are found is Aïr, in the ancient city of Assodé, now in ruins. More commonly it has been noticed a combination of earth and stone is used,

in fluctuating proportions never exceeding 50 per cent, and the stone is used in small cuts and rough (Mester de Parajd 1988).

32 In this case, the hut has no distinction between wall and roof: the plan is circular, and the frame is made up of flexible poles, stuck in the ground, curved and tied to the top (Arecchi 1991).

33 In the present text, the term "banco" indicates a mixture of sand, clay, chopped straw and, sometimes, animal excrement or termite earth.

34 The Haussa traders used to build the houses in the traditional styles of their country. This architectural typology enjoyed double prestige for those ethnics (Djerma and Kanouri) who used to live in huts: it was the home of the rich merchant and resembled western houses. The wealthiest citizens, as the village's leaders, also brought Haussa artisans from far away to build their homes (Mester de Parajd 1988).

35 It stresses, however, that the thermal comfort is minor compared to the huts in humid months, due to the small openings. This problem worsens towards southern Niger where the rainy season is most extended (Mester de Parajd 1988).

36 The raw earth lasts between 10 and 12 years, but only with constant maintenance after the rainy season (Mester de Parajd 1988; Houben, Guillaud 2006).

37 Using the wood as a building material has drawbacks in the Sahel area because termites can easily attack it. Today, the wood is usually treated against the attack of these animals, and the use of palm beams is preferred because they are more resistant.

38 The use of barrel vault has become a common typology in the recent constructions, also in development projects: e.g. the Dosso school built by the Luxembourg Development Program, the Center de Santé Intégré de Garagoumsa in Zinder, the Center de Formation et Developement Communautaire de Balleyara a Filinguè (Abbado et al. 2015: 52–55).

39 Precedents of Haussa vaults can be found in several historic buildings, such as the Palais de Djermakoye in Dosso (Abbado et al. 2015: 18–19), Bomou Palace in Tahoua, the houses in Zongo and the historic districts of Birni in Zinder.

40 The ancient tradition of fresco painted plaster can also be seen among the decorations. The African mural usually plays on contrasts of black and white, ochre and red, but also blue. Examples can be seen in several locations, such as the Maison du Chef or Palais du Sultan in Zinder, Tahoua, Maradi and in the Palais du Djermaloye in Dosso, built in 1713 (Abbado et al. 2015).

41 The finish of the plaster coating can be applied by hand by the bricklayer or with traditional tools (e.g. spatula, trowel) (Houben, Guillaud 2006: 1211).

42 "Néré" is the West African name for the African locust bean, a perennial deciduous tree of the family *Fabaceae*, which is found in a wide range of environments in Africa and is primarily grown for its pods that contain both a sweet pulp and valuable seeds.

43 These decorations are located in the Haussa and Zinder countries, in the Sultan's Palace (Birni), in the Butcher's House (Agadez) and in some painted houses in Dosso, Maradi and Niamey.

Fig. 02.23
Salt lake between Dosso and Gaya, Niger
© AT 2016

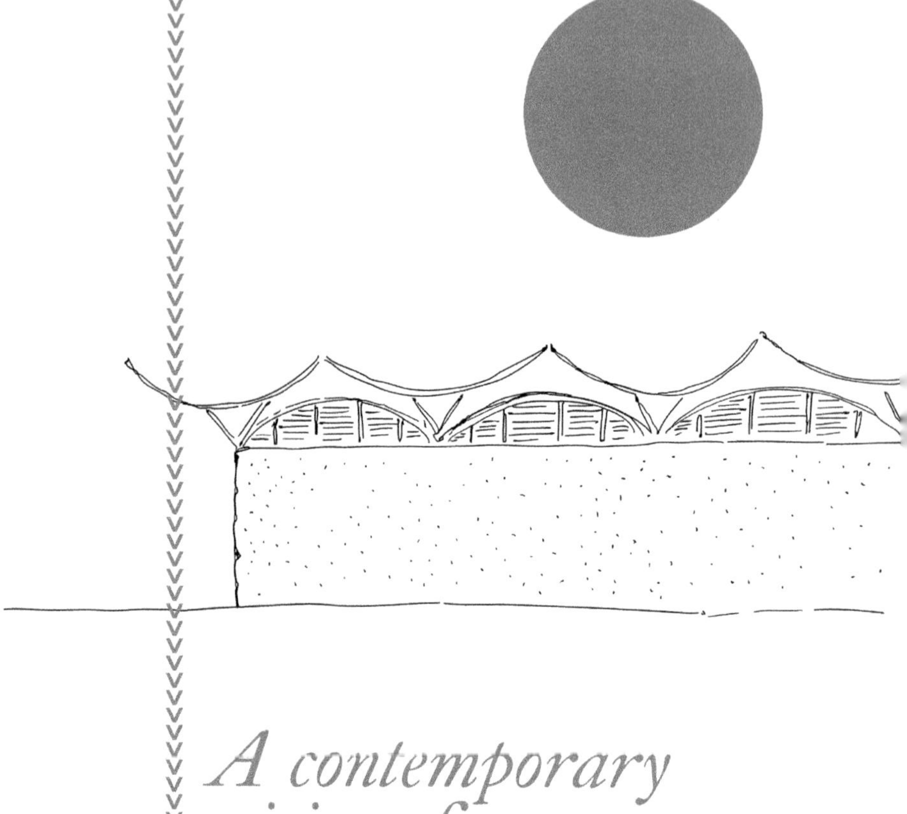

A contemporary vision of raw earth architecture

03

Luca Trabattoni

DOI: 10.1201/9781003157632-03

Architecture is a language. It talks to architects and even ordinary people. It speaks through forms, through space and unquestionably through matter (Espuelas 2009: 33).[1]

The matter of architecture, the construction material, the masonry veneer wall, the finishes are often the first moment of contact between the constructed object and its real users (not the architects, but those who will use the space, who will inhabit the home, who will live in the place).

The appearance (shape) undoubtedly has a great visual impact, and in recent years the world of architecture has been influenced by the needs of the market and society and has pushed towards formal research dictated by the need to amaze. The affirmation of the figure of the "archistar" (Lo Ricco, Micheli 2003) as a reference icon of a given social context has heavily influenced the tectonic value of architecture to the point of bringing the "material" to lose physicality, becoming tinsel, adapted to complex envelopes, interchangeable and not necessary.

In some cases, however, the covering material of the architecture still manifests a personal style; important firms such as those of the Swiss architects Herzog & de Meuron or Peter Zumthor have made material research, often risky and innovative, one of the founding elements of their stylistic cipher. The Dominus Winery, the winery designed in Napa Valley (Betsky 1998), is communicated, for example, through an endless wall of caged rocks in a steel net. The almost raw, "unworked" material is the link with nature, the manifesto of an integration project in the man-made landscape. Looking from outside, the architecture does not disappear, but the presence of the artificial volume is disguised behind the weight of the natural matter.

In contrast, St Benedict's Chapel stands out in the snowy landscape with its iridescent coating. Wood shingles change the colour tone in the lower parts of the building due to a principle of natural oxidation. Applied to the curved shape, the change is homogeneous and controlled.

The material speaks of the ability of the technique, of the ability to work taking natural principles into consideration. The built architecture speaks of the teaching of the artisan tradition applied to a new formal choice. In St Benedict's Chapel the composition of several small wooden elements gives the impression of the technique being controlled. The scandole[2] (wood shingles) are applied one overlapping the other to the vertical facade, to avoid the passage of water. This wall, built with several small elements, has the effect of a pointillist image. The reaction of the material to natural light determines the character of the architectural skin.

The research on the material, as a linguistic application of architecture, has ancient origins, although perhaps the greatest moment of detachment from "traditional" building was that linked to the use of reinforced concrete.

August Perret's Rue Franklin apartment block represents an iconological approach to the material, which adapts itself to "classic" forms and generates a language of estrangement and avant-garde.

"It is a perfectly preserved and highly inspiring work. The framework of beams and pillars gives life to elegantly decorated cladding panels which, thus framed, acquire dynamic rhythm and penetrability" (Zevi 1994).

With Perret and, ultimately, with Le Corbusier's concrete works, which somehow inspired the brutalist period[3] (Banham 1966), the value of the coating material was intended to be an element of language, a "manifesation" of a certain approach to progress and the idea of cutting-edge construction.

Despite the lack of success among ordinary people, brutalism was a "dramatic" movement precisely because it was able to turn the material, in its crudest expressiveness, as a catalyst element of the aesthetic act. Material and structure merge, for example, in the Marchiondi Institute of Vittoriano Viganò, to the point of conveying, with their expressiveness, the formal definition of architecture, according to the emotional strength of its materiality.

Just as the tidy network of wooden planks impressed in the amorphous vaults of the Ronchamp Chapel or in the Church of Notre Dame de Raincy can communicate, so too can the iron panels of the house of Albert Frey or the bricks of the church of Bjoerkhagen by Sigurd Lewerentz.

The material, when used in a clear (explicit) way, is loaded with content that goes beyond the compositional project of architecture. It tells of the constructive research, tells of the compositional principle of the elements of the wall, tells of the desire to experiment and, at the same time, in a pleonastic way, tells of the desire to express a strong language that is heard by all.

After being hired by Lawrence Kochner to design the Aluminaire House,[4] Albert Frey designed a series

Fig. 03.01
Peter Zumthor
St Benedict's Chapel

Fig. 03.02
Herzog & de Meuron
Dominus Winery

of houses in Palm Spring where he carried out experiments on the use of metal sheet and steel structures (Cornell, Lima 2017). Among these, the Frey House, in both its two phases, represents an interesting example of the highly expressive use of metallic material. The corrugated sheet seems almost roughly supported, seen with the eyes of today, it looks more like the barracks of informal settlements than the covering of a villa immersed in the desert landscape of Palm Spring.

Frey's interest in aluminum design, recognized as an emblematic material of modernity, led him to create houses, the appearance of which follows the current low-tech iconography. The Frey house 1, the Frey house 2 and the Loewy House are important precursors of an approach to the use of aluminum and iron intended as noble materials. It is not difficult to combine these projects with those of the Australian Glenn Murcutt or the French Lacaton and Vassal, who will adopt this same material as emblematic of their "idea" of architecture.

The incredible loquacity of the continuous series of bricks of Saint Mark's Church makes Lewerentz's work an important example of the approach to the material of architecture. Widely and extensively described in numerous essays, the obsession that leads Lewerentz not to break or cut any brick generates an almost artisanal building. "A group of friends, that's what the bricks of these two churches look like. [...] The brick used obsessively for walls, vaults and floors is not for him (Lewerentz) a standard element installed with the usual claim of repetitive accuracy [...] In those walls the bricks are not the soldiers of an army, but the ballet dancers" (Espuelas 2009:66). The sequence of the inhomogeneous elements[5] as well as the space between them, which changes according to the particular needs, serve to speak of the constructive act as a moment of devotional love, attention and care.

Unfortunately, with the advances of knowledge and technological potential, "matter" has lost weight. Both the post-modern and the high-tech approaches have proposed strategies for the use of material that is less tied to the construction act, or its physicality, and is therefore lighter and more independent.

In the Centre Pompidou of Renzo Piano and Richard Rogers the material seems almost unseen, it is so crushed by the expressive force of the technological and constructive elements: the tubes, the stairs, the structure, everything has a leading role, and most of the vertical walls are made of glass, ephemeral and transparent.

In the Vanna Venturi House, the material again disappears, not

Fig. 03.03
Sigurd Lewerentz
St Mark's Church

Fig. 03.04
Albert Frey
Frey House

invisible but crushed in some way by the almost iconographic presence of architectural elements: the interrupted tympanum, the square window, the entablature, the ribbon window, the indirect entrance, the fronts all different, the double-pitched roof, the large skylight, the arched window ... too much noise to hear even the material that is hidden under a homogeneous layer of plaster.

Even today, the ephemerality of the covering material, which is almost always distinct from that of structural construction, characterizes most of the buildings. The symbolic value of the material works better than its constructive meaning. Both in the titanium zinc of the deconstructivism (1967) of the works of Frank Gehry and Zaha Hadid, and in the technological efficiency research: "The objects of the most recent generation appear to us more and more often so that we can perhaps say what they look like, but we cannot really say what they are. [...] in the age in which techno-science manipulates the extremely small and manages the vastly complex, on the scale of our perceptions matter no longer appears as a series of given materials, but rather as a continuum of possibilities. Performance and image quality seem to combine with each other in the most disparate ways, giving rise to unclassifiable final solutions. [...] The most durable faux wood, the lighter faux stone, the cheaper faux brick. A world of nameless materials is taking shape. [...] This relative autonomy of the image from material reality is not so surprising: the appearance, in the current technical and cultural phase, has generally become the only reality to which we can refer. In fact, one of the certainties on which modern thought had built up, the one that affirmed the existence of an authentic image of materials, entered into crisis" (Manzini 1986: 33–36).

An authentic image of materials means consistency between the use and aesthetics, the appearance and the "weight" of the materials.

If it is true that almost all contemporary material solutions push towards this "lightening" in function of a better adaptation to the form, it is also true that "heavy" materials remain difficult to lighten.

It must also be recognized that there is a world, and therefore a market far from the common market, where the relationship between the language of matter and construction remains. This is the world of low tech, low technological content and the use of appropriate technologies (Schumacher 1973)[6] and "natural" materials. This architectural approach to technology is often encountered in uncommon contexts, such as those of emergency and cooperation.

In these contexts, the lack of resources, physical and economic, forces architects to find syncretic solutions, where the ontology of the material according to its constructability is a necessity of sustainability.

One of the most common materials in these contexts is raw earth. It is the resource par excellence, always present, even if with different characteristics, always usable, even if with different techniques.

Contemporary

The vernacular scene sees examples of houses made of raw earth all over the world.

Whether they are in pisè (stratified pressed earth built with a formwork system) or in adobe (dried, but not cooked, earthen bricks), or in cob (lumps of raw earth and fibres) or torchis (raw earth spread on a wooden structure), or quincha, greb, bauge ... Almost every geographical region has an example of the use of raw earth as a building material, from Italy to the spectacular cathedrals in Mali.

The most famous "modern" architect who rediscovered the value of raw earth as a building material, economic and practical, and applied it to newly constructed contexts is undoubtedly Hassan Fathy.

In his research work and rediscovery of the Nubian vault and in his work of building New Gourna in Egypt, he always used clay bricks (Fathy 1969). The incredible cheapness of the material, together with the possibility of self-production and the use of a technique, that of the Nubian vault, which can be developed in self-construction, made the New Gourna experience a clear example of a specific approach to building in emergency contexts. The themes of sustainability developed according to passive principles (thermal mass, natural ventilation, etc.), participation understood as the involvement of the inhabitants in the construction of their own homes and the theme of the rediscovery of vernacular technologies have become characteristic of a specific method of architectural intervention.

Currently New Gourna presents several maintenance problems, one of the sayings of the local inhabitants is "rebuild with the Hassan Fathy spirit but not with the Hassan Fathy mud". The problem is related to the chemical composition of the earth used, it was excessively saline and has deteriorated too quickly.[7]

These two aspects, the choice of the technique according to the material and the chemical composition of the material itself are at the basis of the design possibilities of raw earth. Depending on the context, the resources available, the actual composition of the land it is possible to apply different techniques or obtain different results.

The building of the United4Design group in Niamey represents the typical appearance of a raw earth modern building.[8] The compressed earth block (CEB) wall is plastered and appears as a massive straight line. The high and massive wall is covered by homogeneous plaster that reminiscent the colour of the earth and conceals the construction system speaking only of the material. The direct reference to vernacular constructions in clay is evident; just think of the great Malian mosques or the city of Djennè. In earth constructions the final plastering layer serves as protection and is subject to continuous maintenance.

In Caravatti's works the plastic value of the wall is a characterizing element (Caravatti 2006). The vertical surfaces of the Jigiya So Rehabilitation Center, located in Katì, 20 km from the capital Bamako, openly interact with the horizontal surfaces of the paths, defining a panorama of "earth" (Oddo 2015). The landscape defined by the material characterizes a setting, welcomes and separates, determining its own "heavy" presence. Likewise, in Katy Cokò's "neighborhood library" the value of the wall as a line of separation and acceptance is emphasized by its monolithic being. The wall is the emblem of a threshold and being made of earth in a context of "earth", even before the building, it becomes a landscape and guides the gaze and the use of space.

To emphasize the value of this material presence, the covers often break off. Lighter and less "natural", achieved with double ventilated roofs made of metal sheets: they are elements that are different from the wall which find a compositional balance precisely in the distance they can establish from the ground attachment.

"Gando, in Burkina Faso, is the experimental laboratory of Diébédo Francis Kéré. Here he was born and here he returns, every year, to share with his people notions learned elsewhere. In his African buildings, he interprets this wealth of knowledge in respect of a culture deeply rooted in the territory" (Kunsmann 2012: 37). The Gando Campus, where Kerè designed several classrooms and the library, is repeatedly marked by the presence of the brick. The Library or Book Preservation Room (Kunsmann 2012) is a unique example.

The library building forms a physical connection between the Primary School and its extension, sheltering the schoolyard from dusty eastern winds. Keeping with the same material style as the surrounding buildings, the walls of the Library are built with CEBs made with local clay.

This variation of the usual adobe bricks gives Kerè the opportunity to leave the wall without plastering. The lines of brick define the language of the architecture, talking about the construction and the integration with the context (Sinclair 2006: 250). However, the geometry of the library is formally distinct from the others buildings, assuming an elliptical shape. "Space is meant to unite traditional teaching methods between elders and children with the schools' standardized learning environment. In addition to supporting the educational needs of the children of Gando, the Library is also intended as a resource centre for the village as a whole. The Library provides a substantially improved environment for the transfer of knowledge".[9]

The School Library ceiling is the other important particularity of this building: locally produced earthenware pots are used as a skylight to bring the indirect sunlight inside the space.

Traditionally hand built by the women of the village, the clay pots were sawed in half and then cast into the ceiling. These circular openings create a playful pattern and introduce natural light and passive ventilation inside the Library. An overhanging corrugated iron roof sits above this ceiling, protecting the interior and surrounding spaces from sun and rain. The "stack effect"[10] created by the hot metal surface draws cooler air in from the windows and out through the perforations in the ceiling. This provides a passive cooling strategy without the use of electricity.

In the Library of Muyinga, designed by the BC Architects in Burundi the constructive system of the "brick" is pushed so far as to realize a line of columns. "The general form of the library is the result of structural logic, derived on one hand from the material choice (Compressed Earth Blocks masonry and baked clay roof tiles). The locally produced roof tiles were considerably heavier than imported corrugated iron sheets. This inspired the structural system of closely spaced columns at 1m30 intervals, which also act as buttresses for the high walls of the library. This rhythmic repetition of columns is a recognizable feature of the building, on the outside as well as on the inside".[11]

The building talks about the process of construction, which is also part of the process of participation. The building talks using the bricks as words. Luis Khan is used to saying: "if you think of brick, for instance, and you say to brick – what do you want, brick? – and brick says – I like an arch – and you say to brick – look, arches are expensive, and I can use a concrete lintel over you; what do you think of that, brick? – brick says – I like an arch" (Higgins 2009:31).

The adobe (rammed earth brick) asks for the same identity as the one suggested by Luis Khan, but it doesn't

Fig. 03.05
Michael Reynolds.
Global model Earthship Taos

Fig. 03.06
Francis Kéré
Library in Gando

need to become "an arc", because in its own nature, it is the building itself.

The pisé technique allows the wall to speak for itself. A wall almost 80 metres long and at least 3 metres high, defines the image of the Nk'Mip Desert Cultural Centre, in Osoyoos, Canada.

Coloured additives are added to the horizontal layers to emphasize the constructive principle of the wall. The lines, in fact, correspond to the laying phases of the earth, which is pressed and then left to settle. The earth has different colours based on its chemical composition, and using different soils, even though they come from the same region, you can get this kind of "decorative" effect.

A "Loos like" decoration (Loos 1908), however, is not delegated to artistic artifices or tinsel, but determined by the properties of the same material. The building system becomes aesthetics and language.

The same happens in the weekend Ajijic House, in Mexico, designed by Tatiana Bilbao. Here the wall supports not only itself, but also the reinforced concrete roof.

All walls are made from local soil. Due to the necessary earthquake resistance, they are 60 cm thick and stabilized by 8 per cent of cement. The roof is made of 20cm thick concrete slabs.

In the Rosie Joe House, designed by the Designbluff[12] in Salt Lake City, the solid earthen wall is loaded with additional value. One of the fundamental bases of the project is the energy efficiency obtained according to passive systems. The orientation of the plant is the first of these systems. The house is therefore found to have one main front and a secondary front and to develop according to this orientation, therefore in a long and narrow way. As already happens in Michael Reynolds' earthships,[13] the most exposed side functions as a heat storage system, with a long glass wall. The earthen wall in the Rosie Joe House is a perfect passive technical backbone: on the one hand, it functions as structural support for the "impluvium" double-pitched roof (a system that also allows rainwater to be collected); on the other hand, it acts as a storage wall, located behind the glass wall, which completes the passive operation of temperature control; finally, it should be read as a founding element of the space and form of architecture: internally it divides the space, and externally it signs the shape. In fact, it is not only the roof, as already mentioned, that leans on the wall, but it seems like the whole house has "happened" following the earthen wall. As if it were an untouchable pre-existence, the wall comes out of the house, or the house does not reach the end of the wall and appears with its coloured stratification to remember the importance of the settlement.

One of the European masters of earthen construction is undoubtedly Martin Rauch.[14] "Rauch discovered clay architecture not through architecture, but through his education and his first projects as a ceramicist, oven builder and sculptor. His tendency towards applied craftsmanship, towards the thoroughly artistic autonomy of living and environmental design, was already predetermined by his family's farming roots in Vorarlberg, although the decisive impulses for Rauch came to him abroad in that like some of his older siblings he worked for many months as an aid volunteer in Africa" (Kapfinger 2010: 60).

Rauch House, in Schlins, Austria, is a square volume of sedimented dolomite rock. Using the local ground (fully provided by the site excavation pit) Rauch[15] creates a modern house, set into the mountain. The 45 cm walls have an internal insulating layer made of bamboo straw which are and in turn covered with a 3 cm layer of earth plaster.

Externally the house interprets the constraints of the technique and turns them into an expressive instrument. The massive wall has a homogeneous and monolithic tone. One of the problems of raw earth, especially concerning the local climatic conditions, is related to water run-off. Rauch inserts containment lines in the wall during construction, made with stone elements that protrude a few centimetres from the edge of the wall.

In this way the massive image of the house is articulated and "vibrates" using light and shadow. The lines of shadow, like constructive lines, mark the rhythm of the façade, determining the position of the openings and the volumetric waste.

The construction system becomes language not only in its effective rendering, the massive wall of earth, but also by making explicit its virtual projection, the compositional volumetric grid. It is not only the material that is the language of architecture, but the constructive principle of becoming aesthetic.

Even earthbags are a technique that uses raw earth. In their updated version (after military uses) they appear as a sum of igloos. The Ecodomes were designed and developed by Nader Kalili in his research activity for the University of California.[16] The containing element, the disposable formwork represented by the "bag", however, besides being the fundamental point of this technology, is also its weak point. If, thanks to this system, it is possible to create self-supporting structures from the ground to the roof (Kalili 1996), it is also true that the earthbags "suffer" if left exposed to the elements. It is therefore necessary to coat the wall with a thick layer of plaster the components of which vary. Finally, we obtain a building on the ground where the earth is hidden.

Fig. 03.07
Dialog Architects
Nk'Mip Desert Cultural Centre

Fig. 03.08
Anna Heringer
Meti School

The blue dome designed by Jose Andres Vallejo is symptomatic of this misconception. While maintaining all the positive properties of building with the earth, and indeed bringing the additional advantage of the reduced quantity of water,[17] in terms of architectural language the earthbags disappear behind the mass and behind the colour. Constructed in Bogota in 2011, La Casa Vergara uses earthbags to build two kinds of volume: one built as a common igloo, and the other as a flat roof building. Both of them are realized with the same technique, both of them are completely covered with blue plaster. So, the final images of the house hide the technology. It remains, as memory foam, the roughness of the surfaces, smooth and with the kind of imperfections typical of this kind of constructive system.

The school of earth, or Terra dei Bambini,[18] by ARCò represented another degree of ambiguity linked to earthbags. Built within the Gaza Strip in Palestine, the kindergarden mixes multiple constructive solutions to confront a substantial lack of resources. The main walls are made with filled sacks using the excavation site. Developing in a straight line, and thus denying the static principle by form, a system of buttresses is inserted, again using earthbags. The walls of the interior elevation are made of raw earth bricks. Above the walls the double sheet metal cover seems to fly. To emphasize this distance and to obtain the final image of the school, work is done on external plastering between the roof and the wall. Using a mixture of local earth and lime, the exact pigmentation is sought, so that the building blends in with the landscape. The earthen building is dressed in earth to look as if that is what it really is.

The contemporaneity of raw earth as a material not only efficient and economical, but as a vehicle of architectural language can also be seen in the experiments related to its technology.

Two workers next to a wall that seems to have been built off the scale. The unitary megablocks mounted as bricks are prefabricated earthen elements. The Ricola Herb Centre in Laufen (Basel), Switzerland was designed in 2012 by renowned architects Herzog & de Meuron with a facade constructed by master clay builder Martin Rauch, the building is a high-volume long building with a flat roof and façade built using raw earth. Façade elements made of compacted local clay sourced from the Laufen valley will form Europe's largest loam building in 2014. The prefabricated earth elements were compacted into a formwork in a nearby factory and stacked on site to form walls. Thanks to the plasticity of the loam, the seams can be retouched, giving the overall structure

Fig. 03.09
Nader Khalili
Usual earthbag
building

Fig. 03.10
ARCò
Kindergarden in Gaza Strip

a homogeneous appearance.

"Ever the material experimenters, Herzog & de Meuron chose to prefabricate panels of rammed earth in a nearby factory and have them hoisted into place by crane, rather than construct the material in situ, as is typically done. The architects also chose to incorporate lime mortar and volcanic tuff into every eighth layer of the material as a way of preventing erosion. There is an obvious irony in the fact that one of the world's most materially innovative design practices has selected one of the oldest and most common materials for its latest project. However, given the relative ignorance about rammed earth in many parts of the world today, the Kräuterzentrum serves as a welcome and inspiring example of a more environmentally sensitive architecture"[19] (Brownell 2014).

The wasp potter (Sceliphron spirifex) is a hymenopteran belonging to the family of *sphecids* (*Sphecidae*), endemic to many areas of northern Africa and southern Europe. The nest is composed of mud, which is expertly modelled and cemented using a particular substance produced by the animal's mouth. The mud nest is composed of a series of cells, sometimes cylindrical, which form a resistant structure that can reach almost the size of a fist.

"The Vespa Vasaia is our development model, a perfect approach to build low-cost houses, with natural materials, at zero kilometres". Says Massimo Moretti, founder of WASP, the company that created Gaia, the first raw earth house printed with a 3D printer. A metal arm rotates suspended on a metal lattice frame. When it passes, it builds a hollow honeycomb wall. A mixture of chopped rice straw and rice hybrid will be poured into the cavities of the wall. The wooden roof is supported by a series of wooden pillars.

The formal constraints are high, and the limits to date are evident, in the dimensions in height, in the usability of the wall. The wall is not structural, but only a wall. In addition there are the usual limits of earthen buildings, mostly related to maintenance. The advantages are also declared. The rapid construction (30sqm in ten days) the economy (only €900) and obviously the passive energy efficiency. The bioclimatic project signed by RiceHouse,[20] exploits the passive contribution of the sun thanks to the south-west orientation where a large window is positioned to optimize natural light; also the structure is made of wood, as well as the roof, with insulation in lime-hull (RH300), both in the screed against earth made of lime-hull, which allows it to reach an energy requirement equal to a class A4).

The incredible value of this experimentation, however, is the recognition of the role of the material:

the raw earth[21] is the fundamental ingredient of this building, the one that allows not only its realization, but also its energy performance.

Openings

Sustainability is perhaps today the only parameter of homogeneous evaluation of architecture. Sustainable is in many cases used as a synonym of "quality".

The panorama of protocols that define the sustainability of buildings is varied and covers a very wide range of variables.

From the efficiency of the building package to the implantation performance, from the average consumption to the origin of the materials, from the orientation to the forced mechanical ventilation systems, and others. The object is always and, in any case, and perhaps rightly, a quantification of the performance, not forgetting that the efficiency of a building is only one of the important aspects of its evaluation. It is therefore not difficult to imagine how efficiency can become a screen that obscures all other aspects of architecture; therefore, it becomes inefficient (Snozzi, Merlini 2014) or at least an incomplete judgement. In the incompleteness of the evaluation, the debate on language remains pending. Undoubtedly, speaking today of sustainable buildings, it means talking about the construction process, and therefore about technology and therefore also about materials.

It is not easy to define a language for a specific scientific approach (Pantaleo 2016),[22] and perhaps it is not even necessary. Undoubtedly, however, there are some materials and some technological choices that are more related to a particular line of thought. The low-tech and low-cost approach, that of "appropriate technologies" (Schumacher 1973), considers "bulky" materials and construction systems, whose role in the construction and architectural composition is binding for a formal definition.

Rock gabions, bags of earth, straw bales are the most unusual translation of this approach, which however also has some more conventional "translations". Think again of the works of Lacaton and Vassal, where the choice of low-tech becomes a distinctive feature of a strongly ideological architecture. The buildings of the French architects work with light structures, filter spaces that implement the housing dimension by working passively. The recovery project of the residential block in Bordeaux (Zabalbeascoa 2019), or the residential intervention in Mulhouse are examples of how the low-tech approach can be not only current but also extremely effective.

In Bordeaux, the architects proposed a recovery project for a large residential complex by creating a system of semi-open spaces, which positioned the facades with the right orientation (greater collection of sunlight), manage to be living spaces and technical elements. Acting as bioclimatic greenhouses, in fact, these environments are popular residences, which change their appearance, change their internal space and have the possibility of improving their own energy efficiency, counting on excellent heat sensors. The whole project is made with exposed reinforced concrete, iron structures, polycarbonate panels and curtains. The "rough" aspect of the intervention still manages to give new life to the huge residential complex.

La Cité Manifeste in Mulhouse uses the same components to create a new public housing sector. The project works with modular repetition, through the construction of a system of metal arches resting on a concrete base. In organizing the spaces, all the apartments are equipped with a filtered environment, aesthetically identical to the other domestic environments, but which is not heated and instead works as a heat collector and as an open space. The image of the complex recalls the greenhouse system of plant nurseries. The base in raw concrete, the metal structure and the panels in corrugated polycarbonate and corrugated sheet are the language of this residential complex (Rocca 2010: 131).

Metal and polycarbonate, although aesthetically very recognizable, certainly do not have the physical heaviness of other materials mentioned. Straw, stones and earth are materials inherited from tradition, but which are finding their own consolidated place in the contemporary world.

Straw, for example, beyond the experiences of Barbara Jones, is finding interesting updates. The building of the Archetype Studio,[23] Enterprise Centre at the University of East Anglia, for example, is experimentally presented with a vertical façade completely covered with exposed straw. The system of prefabricated and assembled panels guarantees for the first time the application of this material in a "modern" and visible way. The straw that normally would need to be protected from the weather finds through this magnificent technical solution, the possibility of maximum aesthetic expression.

Just like straw, raw earth has also found a place in contemporary architecture.

The search for a contemporary architectural language that also manifests itself with a specific constructive approach finds in this material a convincing answer.

The plastic potential applied to energy performance makes it a significant resource. In the world of cooperation and emergency (humanitarian, resources, etc.), numerous architectural firms have exploited this material.

But even in less ordinary areas, raw earth is becoming a significant construction opportunity. An important signal linked to this trend is given by the fact that some companies are starting to market construction systems based on this material, which work with the pressed wall, the bricks, the plaster.

Several architectural firms have exploited this material in experiences of cooperation[24] and emergency design (humanitarian, resources, etc.).

>>> Notes

1 "Matter is to architecture as sound is to text. The matter resonates. Severe or acute, bilabial or dental, fricative or guttural. Roland Barthers said – I have a disease, I see the language. – I allow myself to paraphrase it: I have a disease. I listen to the matter" (Espuelas 2009: 33).

2 It is a particular traditional technique used to make the wooden roof. "Scandole" is the Italian name for the small pieces used to finish the roof.

3 "Behind all aspects, the new brutalism [...] lies one undisputed architectural fact: the concrete-work of Le Corbusier's Unité d'Habitation at Marseille. And if there is one single verbal formula that has made the concept of Brutalism admissible in most of the world's western languages, it is that Le Corbusier himself described the concrete work as beton brut" (Banham 1966: 16).

4 Alluminarie House was the futurist project proposed by Kochner and Frey for the the biannual buildings products display of the Allied Arts in 1930. "The design was influenced by the modern movement which had reached its zenith in Europe in terms of both image and its progressive political concern with housing. Albert Frey's experience in Le Corbusier's office provided the principle idea for a 'maison minimum' which had been a common European concern for the housing problem of the industrial revolution city. However, it also seems to have been influenced by the American growth of industrial mass production materials and the advanced representation of this in Buckmister Fuller's work [...] It was assembled in ten days in the Grand Central Palace that was adjacent to Grand Central Station and opened on April 18th, 1931 for one week.

Over 100,000 people toured the house" (Cornell, Lima 2017).

5 Lewerentz works with different brick formats. Furthermore, having to cut the bricks also works on the quantity of mortar, which serves as compensation to reach the necessary dimensions.

6 The concept of Appropriate Technologies elaborated by Schumacher in the book *Small Is beautiful* of 1978, arises from a criticism of the consumerist society which, according to the author, makes an obvious mistake: that of basing its development principle on the use of natural, exhaustible capital, as if it were an income. In the same way Schumacher criticizes another principle of modern society closely connected to the first, namely that of the idea of unlimited technological growth, which as a model, in his opinion, has already proven to be unable to solve the problem of poverty nor to generate sustainable living (Schumacher 1973).

7 "Outreach & Education – Past, Present, and Future of Hassan Fathy's New Gourna, Egypt:" is a documentary promoted by the World Monuments Fund (WMF). New Gourna was placed on the 2010 Watch, and UNESCO and the World Monuments Fund joined forces. UNESCO surveyed buildings and documented the conservation needs while the WMF team interviewed more than a third of New Gourna's 174 households, analysing conditions, identifying changes in the environment and examining the use of space and adaptations to the buildings over time. WMF created a community assessment to integrate social and economic concerns into planning and decision-making about the future of the village and its preservation, published in a report on New Gourna Village that can be found on WMF's website, wmt.org. The data expressed in the documentary are also in the report which is called: New Gourna Village, Conservation and Community – promoted by the World Monument Fund (2011) with the support of Robert W. Wilson Challenge to Conserve our Heritage and in collaboration with the UNESCO World Heritage Centre and the Luxor Governator of Egypt.

8 http://www.united4design.com.

9 www.kere-architecture.com.

10 The stack effect or chimney effect is the movement of air into and out of buildings, chimneys, flue-gas stacks or other containers, resulting from air buoyancy. Buoyancy occurs due to a difference in indoor and outdoor air density resulting from temperature and moisture differences.

11 http://architects.bc-as.org.

12 DesignBuildBLUFF is a studio headed by philosopher and architect Hank Louis. At the University of Utah College of Architecture and Planning first- and second-year students of architecture are given the opportunity every spring to build housing for poor families on the Navajo tribal reserve near Bluff, Utah (http://www.designbuildbluff.org/).

13 Michael Reynolds is a well-known American architect who has invented and built a model of housing completely independent from an energy point of view. Using passive systems and recycled and natural materials, earthships are self-built homes. The orientation, thermal mass, passive heat collection systems are among the systems used by Reynolds (www.earthshipglobal.com).

14 Born in 1958 in Schlins, Austria, Martin Rauch is one of the greatest experts of the technical pisé internationally. Since 2002

he has been honorary professor of UNESCO for the Department of Architecture of Earth.

15 House Rauch was designed by Martin Rauch and Ralph Boltshauser and built by the company they founded Lehm Ton Erde Baukunst GmbH (architectural limited liability company for construction of clay and earth).

16 SuperAdobe is a form of earthbag architecture developed by architect and CalEarth founder Nader Khalili. Using long sandbags ("SuperAdobe Bags"), barbed wire, on-site earth and a few tools, Khalili devised a revolutionary building system that integrates traditional earth architecture with contemporary global safety requirements and passes severe earthquake code tests in California. This technology has been published by NASA, endorsed by the United Nations, featured in countless world media outlets and was awarded the prestigious Aga Khan Award for Architecture in 2004. It comes from years of meditation, hands-on research and development. It was inspired by traditional earth architecture in the deserts of Iran and adapted for modern usage. Simplified so that anyone can build.

17 The earthbags are made by filling the polyethylene bags with dry earth. A layer of barbed wire prevents slippage of the bags, which can potentially also be filled with sand. It is therefore possible to enormously reduce the need for water, intended only for external plastering.

18 Children's land (trad. aut).

19 Blaine Brownell: http://arquitecturasdeterra.blogspot.be.

20 www.ricehouse.it.

21 The composition of the walls of GAIA is a mixture made of 25 per cent local raw earth (30 per cent clay, 40 per cent silt and 30 per cent sand) 40 per cent shredded rice straw, 25 per cent rice husk, 10 per cent hydraulic lime

22 In his book on his practical experience, Raul Pantaleo (Tamassociati studio) talks about an unusual way of seeing beauty: a kind of "dirty beauty" related to the contamination of the architectural idea with the buildings' limits due to the emergency context (Pantaleo 2016).

23 https://www.architype.co.uk/project/the-enterprise-centre-uea/.

24 The Brundtland Report (also known as Our Common Future) is a document published in 1987 by the World Commission on the Environment and Development (WCED) in which, for the first time, the concept of sustainable development was introduced. The name was given by the coordinator Gro Harlem Brundtland, who in that year was president of the WCED and who had commissioned the report. Its definition was as follows: "sustainable development is a development that meets the needs of the present without compromising the possibility of future generations to satisfy their own".

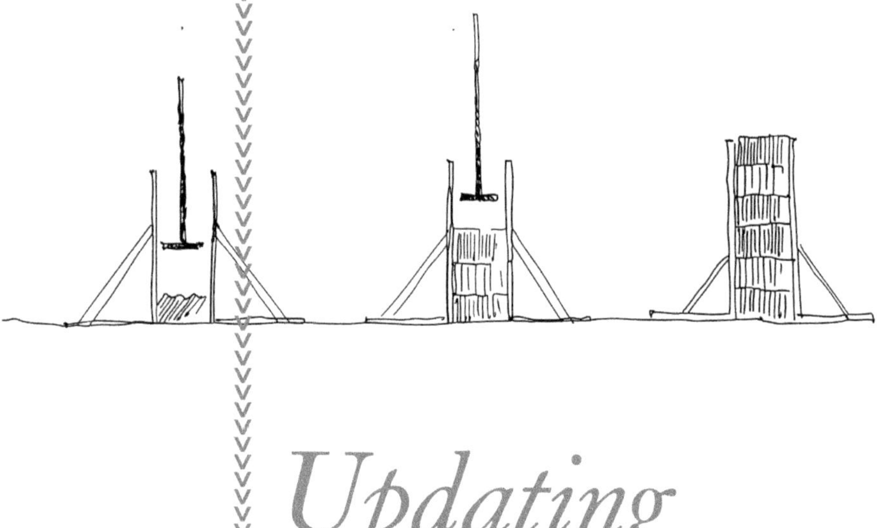

Updating vernacular architecture

04

Luca Trabattoni

DOI: 10.1201/9781003157632-04

The vernacular is by its nature a known matter. Rudofsky called it "non-pedigreed architecture" (Rudofsky 1964), the spontaneous, anonymous, rural, native, indigenous, traditional architecture, that heritage that is part of the art of building that was and is, the cultural heritage of peoples.

Rudofsky's vision seeks architecture not limited to formal language or the definition of a style but rooted in a close relationship with the environmental and cultural context.

As Paul Oliver reiterates in his *Encyclopedia of Vernacular Architecture of the World* (Oliver 1997), the non-monumental, non-formal and un-designed architectural heritage of the world is understandable only through a reading filtered by knowledge. The different people in the world build and live in different forms and to understand these unique and distinct forms of architecture we have to look at them as an integral part of the societies and cultures that produced them. From this point of view, culture and society are specific to the place, clearly delimited and potentially fixed in their nature.

The vernacular architecture represents a category of poster representative buildings of a certain way of life and living. A way linked to the knowledge of the climate, the ability to manage existing resources according to the needs of the community.

The value of vernacular architecture is therefore primarily in its immanence: it is being the emblem of a specific place and a specific culture conditions its construction and its language. Architecture then becomes not only manifest but teaching, an instruction manual to understand the "genius loci" of a place (Norberg Schulz 1979). As Norberg Schulz explained, when a man dwells, he is simultaneously in space and exposed to a specific environmental character. Through the identification with this spirit of a place, the genius loci, a meaningful man-place relationship develops and strengthens the sense of belonging of man.

Vernacular architecture is built with the resources present in a given place to contrast the climatic conditions of that same habitat. Entering the merits of the African continent, and specifically of the Sahel area, think for example of traditional earth huts.

The climate of Niger, like much of the Sahel area, is hot and windy. These huts are mainly built on the ground, with massive walls made of pisé or adobe to guarantee strong heat protection thanks to thermal inertia. The cone-shaped roof made of wood and straw which favours the concentration of hot air at the top contributes to the environmental quality of the internal environment.

It is clear that architecture is indissolubly determined by the constructive act, in a "tectonic" act (Frampton 2001), where the harmony between form, language and matter is complete. Although modest in their appearance, the huts typical of the Niger area represent profound teaching. They include the existing materials, the only ones present in sufficient quantities, and use these materials coherently and completely. The living space is unique, as often happens in the houses in the Sahel area, the relationship with the fence is changeable and not consolidated.

Of course, it is a static language, unprepared for climate and corporate changes, it is a language that needs updating. The need also expressed by Frampton (2001) to think of an architecture taking into consideration both the local culture and the quality of the landscape, implies an ability to read the changes in the landscape and local culture. Frampton's critical regionalism seeks architecture based on a "post-form" balance between the natural environment and the cultural heritage that identifies societies.

The geographical characteristics and cultural heritage will then be decisive for the ecology, climate and symbolic aspect of the place.

However slow it is in its mutations, the African continent also suffers the influence of western society and its progress. The cultural clash generates inconvenience, related to cultural unpreparedness, as Aime teaches us when talking about waste management (Aime 2000). For example, plastic is managed as perishable waste, generating open dumps exposed to the wind and to being ingested by animals, with all the (harmful) consequences that this situation can generate.

The relevance of the study of vernacular knowledge is also emphasized by the reference to current global concerns for sustainability and cultural identity which continue to question the processes of modernization and globalization (Asquith, Vellinga 2006).

It is important at this point, however, to point out that the romantic and Rousseauian vision of the vernacular actually represents a demeaning synthesis of this architecture. Precisely in the face of the indissoluble link with the cultural aspects of local society, vernacular architecture represents a complex system of multiple readings. Among these, the need to update the same techniques is partly also linked to the need not to forget. As he taught in the 1970s Hassan Fathy, one of the great problems of the poor contexts, as is the case for many Africans, it is the loss of memory (Fathy 1969). Some admirable techniques, such as the Nubian vault, tend to disappear, as they are linked to the knowledge of craftsmen who cannot pass on their knowledge.

Furthermore, the perpetual abandonment of the rural dimensions in favour of large metropolises, especially by the younger part of the population (Davis 2006), has led to the integration of artificial materials, typical of the urban context (iron, sheet metal, plastic) with hybridization of local resources.

But perhaps, it is precisely in this dimension that it is possible to find that advancement of the appropriate technique (Schumacher 1973) consistent with an idea of happy degrowth (La Touche 2007) that updates the vernacular one.

Focusing therefore on Niger, where the increase in poverty, especially in suburban and peri-urban contexts (Saxena, Vyas 2019), has meant that the relative cost of natural materials (wood) has increased considerably. In these contexts, as happens in many parts of the world, the last truly democratic resource remains the earth.

Raw earth is the most widespread material in the world. Examples of architectures made with this material can be found everywhere, from Italy to Africa. Raw earth can have different conformations depending on the place and use. The dough used to build is often a mix of lime, earth and water. Often fibres are added to reinforce the seal of the dough.

What we will do in this text, however, is not to linger on the technical and chemical needs of this material, but to investigate its most common uses and what the possible modifications are.

Raw earth has many names, linked to different techniques. Finally, the raw earth is never sufficient to complete the work. In almost all raw earth techniques, with the exception perhaps of the wooden compass and the Nubian vault, the earth must be accompanied. The roof is one of the most difficult elements of the composition to integrate into this technique. But even the walls can have different sizes and different compositions, be integrated with wooden frames, for example, or not.

Building technology

Raw earth is a natural and biocompatible material. Its impact on the environment is minimal, as it is an abundant resource all over the world. It is almost completely recyclable as at the end of its life cycle it can return to nature without any special treatment.

In addition to this basic sustainability issue, raw earth building has got some other interesting features, such as considerable mass and, consequently, thermal inertia. The walls are usually about half a metre thick, they have excellent heat storage capacity and can retain dampness in the air. Both heat and humidity are released into the environment by themselves when necessary. Even the soundproofing properties are good, also thanks to the porosity of clay. It's easy to understand why fire resistance is also positive.

From a constructive point of view, raw earth walls have good strength and excellent capacity, moreover, clay is a pliable, ductile and a particularly binding material. These properties make it suitable for different shapes and sizes. Even if the structural resistance of a rammed earth wall is limited, with the necessary precautions this construction technique can also be applied to tall buildings, such as the Great Mosque of Djenné.

Among its negative sides, we can point out its sensitivity to water, which might compromise its performance and strength, if not employed with adequate attention. That is why it is always necessary to evaluate a coating, for example using lime, which protects the walls from atmospheric agents. On the other hand, it is advisable to set up some horizontal barriers in order to counter the rising damp.

Raw earth construction can be traced back to the origins of living in dwellings, it is not only typical of rural buildings but there are also many examples of it in the cities. Earth construction techniques have been known for over 9000 years. Mudbrick houses dating from 8000 to 6000 BC have been discovered in Russian Turkestan (Pumpelly 1908). Rammed earth foundation dating from ca. 5000 BC have been discovered in Assyria. Earth was used as a building material in all ancient cultures, not only in homes but for religious building too (Minke 2006).

One of the oldest buildings in pisé can be found in Mehgrah, a city in the Indus valley in Pakistan and in the archaeological excavations of Carthage, with buildings in clay and adobe even six floors high (second century AD).

Plinius wrote that earth walls were built in Africa and Spain and that in Asia buildings were made with dried bricks (Agricola 2004). In China, however, after 581 AD, they began to use pisé construction and many parts of the Great Wall were built with rammed earth before being covered with stone.

The earth, as a building material, can be applied to several techniques that used in addition other structural materials such as a wooden structure or straw fibres or another kind of framework. Amongst other materials wood has good compatibility with earth, because of its reaction to humidity and temperature. There is a good scientific library on this issue, and maybe the best one is the manual written by Gernot Minke (2006). Minke is one of the most important references for bioconstruction. He has realized several prototypes and real projects, experimenting with new and old technology. He has written some important books about new low-tech technology.

To clarify how complex the "raw earth world" could be it's enough to read what Minke says about the nomenclature of it: "when used as a building material, is often given different names. Referred to in scientific name term as loam, it is a mixture of clay, sits (very fine sand), sand, and occasionally larger aggregates such as gravel or stones. When speaking of handmade unbaked bricks, the terms 'mud bricks' or 'adobes' are usually employed; when speaking of compressed unbaked bricks, the term 'soil block' is used. When compacted within the formwork, it is called 'rammed earth'" (Minke 2006: 27).

In this chapter, we want to show some of the most famous techniques that applied this material. We will use one of their names and we will illustrate where possible, some ways by which it is possible to update the technology described or use it to compose the language of architecture.

Due to the high number of variations possible and the widespread use of this type of technology, the following list is only a selected choice of the most common methods that can be used to work with earth and loam.

There are many other ways to use the same material, vernacular and even modern. A fine example, which is not on this list, is the work of Michael Reynolds. In the typical building, the Earthships, Reynolds applies the earth in different ways, such as plaster, as a filling element in a wall of used tyres, as a binding material (mortar) in wall of cans or wall of bottles, to prove the high versatility of this material. Earthship houses are a conceptual living space focused on energy independence. To achieve this, Reynold created an extreme building: each house has a very linear orientation and is developed as a straight/lengthwise building. In this way it obtains two very important sides: one of these is aimed at obtaining the greatest amount of solar radiation possible and is completely realized as a greenhouse. The other side is normally completely closed. It is made against the ground, most often using used tyres as a structure to keep the soil dry inside. In this way, Reynolds obtains a building that works passively for air conditioning. The other walls are mainly made of waste materials and raw earth.

Earthship is an excellent example of how it is possible to manage natural material and recycled waste material to obtain, by applying passive sustainability principles, a very self-sufficient house.

>>>>>>>>>>>>>>>>>>

Adobe

Adobe is a mixture of clay, sand and sun-dried straw used by many populations in every age to build bricks.

Adobe bricks or mud bricks are a mixture of sand, clay, water and some organic reinforcement such as straw or sticks. They are not difficult to realize, and constituent ingredients can be found almost anywhere in the world.

Building with the earthen block is widespread in all hot-dry, subtropical and moderate climates. Earth block buildings dating from 8000 to 6000 BC have been found in Turkestan (Pumpelly 1908), and ones from 4000 BC in Assyria. You can find them in the south-eastern dry areas of Spain, in Italy, but particularly in Sicily and Sardinia, in semi-desert areas of Africa and Central America.

Earthen block buildings can be made by throwing wet earth into a formwork. This is why adobe is a low-cost, readily available construction material, usually manufactured by local communities. Once made, the wet mixture is pressed into moulds and left to dry in the sun. Different kinds of mould can be used, unusually they are made of timber. It can be removed from the moulds quickly to free them and continue the brickmaking. In this way, bricks can be made faster and building can progress rapidly. However, the bricks do not have to be excessively dry, and consequently, they are wrapped in an envelope, often made of horsehair. Drying bricks in direct sunshine in hot, dry climates can cause cracking, so it's better to let them dry in a shady place. Adobe bricks are easy to make, don't require any heavy machinery and provide versatility in design and construction due to their unit size (Austin 1984). The size of a single brick must be correct so that the worker can handle it with only one hand. Mud mortar is typically used between the blocks.

The great thermal peculiarity of this technique allows it to retain heat in winter and release it in summer, maintaining thermal comfort in every season (Michael et al. 2016). However continuous maintenance is required, indeed rain is its biggest enemy, so layers of mud are necessary to protect it.

The highest building realized in adobe is the Great Mosque of Djenné, in Mali (Marchand 2009). Its appearance is characterized by the sequence of wooden poles that

Fig. 04.01
Adobe

Fig. 04.02
Francis Kéré
CEB Gando School

emerge from the earthen wall and which are used for the maintenance phases that also take place twice a year. Even if a building is built with an ancient version of the adobe brick, it preserves the maintenance problems typical of this technique. The Djenné-ferey system is the traditional construction method, consisting of an overlapping of balls of raw earth still wet, which covers the role of both brick and binder. This technique differs from that of toubabou-ferey, introduced by Westerners, with the manufacture of dried bricks, held together by raw soil still wet as a binder.

Recently, adobe technology has been updated with the realization of CEB. This way of making bricks combines the mixture of the earth with other materials, such as lime, aggregates and even cement. As 80 to 90 per cent of the volume of a CEB is soil, then soil selection is still the most important part of formulating CEB mixes. If the blocks are stabilized with a chemical binder such as Portland cement, they are called compressed stabilized earth blocks (CSEBs) or stabilized earth block (SEB). Typically, the original soil volume is reduced by about half.

The advantage of adding aggregates and cement (normally not more than 8 per cent) to the earth, is that it gives the brick greater resistance. To make CEBs it is necessary to have a special machine, a pneumatically operated press. The CEB is far more resistant than the normal adobe to the wind and rainy water erosion. The South African Department of Water Affairs and Forestry considers that CEB is an appropriate technology for a developing country, as are adobe, rammed earth and cob.

In 2002 the International Institute for Energy Conservation was one of the winners of a World Bank Development Marketplace Award for a project to make an energy-efficient Dutch brick-making machine for home construction in South Africa. By making cheaper bricks that use earth, the project would reduce housing costs while stimulating the building industry.[1] The machine would be mobile, allowing bricks to be made locally from earth.[2]

The CEB system has many advantages. On-site materials can be used, which reduces cost, minimizes shipping costs for materials and increases efficiency and sustainability. The wait-time required to obtain materials is minimal, because once the blocks are pressed, materials are available after a short drying period. The uniformity of the blocks simplifies construction, and minimizes or eliminates the need for mortar, thus reducing both the labour and materials costs. The blocks are strong, stable, water-resistant and long-lasting. CEBs can be pressed from damp earth. Because it is not wet, the drying time is much shorter. A single mechanical press can produce from 800 to over 5,000 blocks per day. CEBs can be manufactured to a predictable size and have true flat sides and 90-degree angle edges. This makes design and costing easier. The CEBs are insect resistant because the walls are solid and very dense, and have no food value (Garg et al. 2014)

Francis Kéré architecture is a good example of the use of this type of brick on the ground. Brick gives a texture to the wall, but also maintains the hue of the ground.

Since most of the elements are made of earth, CEBs maintain the energy performance of adobe. Furthermore, in small dimensions, they can also be used as a structural element even if most of the time they are just used as non-structural masonry.

>>>>>>>>>>>>>>>>>>>>>>

Pisé

The construction technique of "pisé" ("pressed", in French) or rammed earth is based on building walls with slightly damp earth (to avoid cracks as it dries) compacted with special tools of a limited size and removable wooden formworks, to allow their movement.

Clay mass can be sometimes lightened by adding chopped straw and dried grass. In dry areas pisé is set on levelled and clean ground. On the contrary, in humid areas, we use a foundation made of bricks or rough stones bound with cement mortar.

The wall can have a variable thickness and is made of layers. You can bat clay with wooden instruments which have different shapes, for example, a large head or a sharp point (or modern mechanical instruments), in order to make it compact. In this way, it gets less friable on the surface and has better mechanical qualities. When the earth has hardened, the formwork is removed to re-arm it on its side, maintaining some contact with what was made before. This procedure is repeated until the desired height is obtained.

Frequently the corners of the buildings are made with crossed casting, reinforced with layers of lime, bricks or stones.

The pisé is a very ancient technique of which there are testimonies – made with almost identical building principles – practically all over the world.

The geographical spread and the popularity over time of the pisé – which demonstrates its relevance and versatility – can be traced back to at least three factors.

First of all, it is easy to find a type of soil suitable for the pisé construction; secondly, the smaller quantity of water, compared to other construction techniques on the ground necessary

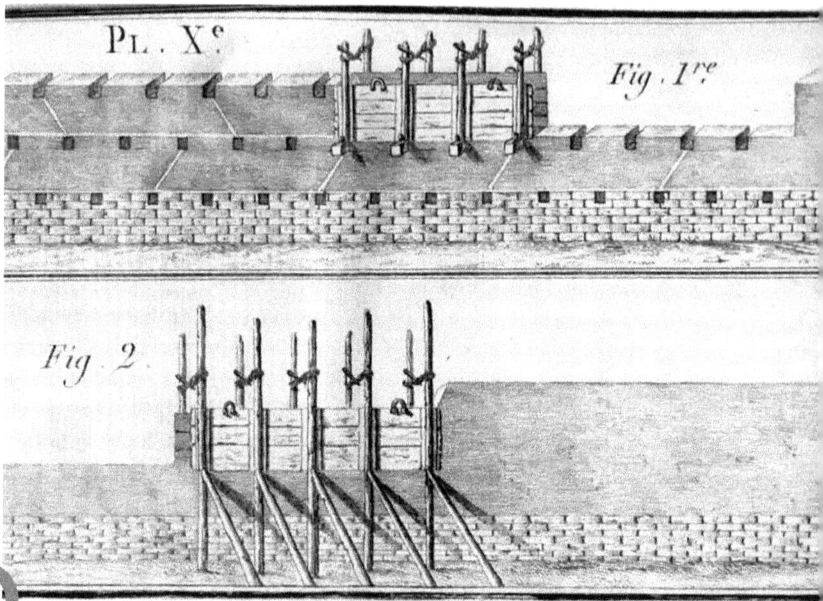

Fig. 04.03
Pisé wall construction drawing by François Cointeraux (1791)

Fig. 04.04
Alternative formwork from a student's maquette
[Sabatini, Ciurlo, Cabaud]
"Necessary Architecture Workshop".
Politecnico di Milano
February 2019

for its preparation; finally, the lack of dependence on the availability of timber.

The real disadvantage of this technique is related to the need to use formwork. Normally these structures are made with wooden elements, a material that is not always easy to recover and often very expensive. The realization times are also quite long: the pisé provides a layer by layer construction. The levels are very small and must settle for several days before becoming solid. Finally, compared to other techniques, the use of water is fundamental.

The walls in pisé can be structural if applied to small buildings. To create the openings in the walls you can proceed traditionally, using wooden or other material lintels that support the empty space.

An interesting way of modifying the language of this technique could be to work on the formwork. This element, which is usually the most expensive part of the process, could be realized with a material that can be re-used for another part of the building, such as the iron sheet used for the roof. By being careful not to make a hole in the iron sheets in the fixing process, it would be possible to realize a wall with a very particular language.

This action could lead to an interesting new vernacularization of the technique; in fact, the type of iron sheet that can be found all over the world changes with the country in which you work. It could be wavy, it could be aligned, it could be square. This would imply a specificity of the wall defined by region, linked to a non-natural material but linked to market supplies.

Another interesting new way of using this technique is related to the decorative use of the earth layer. The building process is closely related to the building timing: to have a good pisé wall it is necessary to respect the sedimentation period of each layer (barely 30 cm high). The SIREWALL system that has been used, for example, in the Nik'Mip Desert Cultural Centre is a contemporary way of using the building "timing" as a process of languages. The horizontal lines that characterize the finished wall are obtained by working on the mixture of the local soil.

The SIREWALL is a structural sandwich wall system typically 18" to 24" thick. The local soils are combined with 6–10 per cent cement and are compacted on either side of a hidden insulation core. The system is stabilized with compacted earth and rebar (rebar is used in seismic zones), with 4" of rigid insulation hidden in the centre of the wall.

>>>>>>>>>>>>>>>>>>>>

Cob

The Cob technique is a very old method of building that uses a mixture of earth and natural fibres with a strongly plastic consistency.

Cob is an English term attested to around the year 1600 for an ancient building material that has been used for building since prehistoric times. The etymology of cob and cobbing[3] is unclear, but in several senses means to beat or strike which is how cob material is applied to a wall.

The mixture of the earth must be very fat, rich in silt and clay, with the addition of straw and fibres. It is quite alike adobe in that the basic mix of clay and sand is the same, but it usually has a higher percentage of long straw fibres.

Traditionally the mixture is worked and kneaded with feet on the ground until it becomes homogeneous; for this reason, it is fairly labour-intensive.

After a day, the dough is worked by hand, then cut into blocks with a spade which can be tossed from one person to another during the building process (Snell, Callahan 2009).

Later it is put into place and stacked in successive courses with an average thickness of 50/60 cm to create walls with irregular blocks of considerable thickness.

When the wall is completed, it is still fresh and is regularized using a spade and its surface is ruled with a rake to make it easier to plaster.

The walls of a cob house are generally 61 cm thick, and windows were correspondingly deep-set, giving the homes a characteristic internal appearance. The thick walls provided excellent thermal mass which was easy to keep warm in winter and cool in summer (Goodhew, Griffiths 2005). Walls with a high thermal mass value act as a thermal buffer inside the home. The material has a long lifespan even in rainy and/or humid climates, provided a tall foundation and large roof overhang are present.

Cob is fireproof, while "fire cob" (cob without straw or fibre) is a refractory material (the same material, essentially, as an unfired common red brick), and, historically, has been used to make chimneys, fireplaces, forges and crucibles. Without fibre, however, cob loses most of its tensile strength. The mixture reaches its solid consistency and complete drying after about 15 days according to the thickness of the wall and the weather conditions.

The consistency of the mixture will not need any formwork to make the blocks and to delimit the masonry. The cob technique, therefore, allows building load-bearing walls, which have the feature of being monolithic. Because of all the straw, cob can be slightly more insulating than adobe. Cob construction is particularly easy

Fig. 04.05
Anna Heringer,
vocational school in
Rudrapur

Fig. 04.06
Anna Heringer,
Meti School

to learn, requires no fancy equipment, uses local materials and can be done in small batches as time allows making it extremely accessible to a wide range of people.

Some of the oldest man-made structures in Afghanistan are composed of rammed earth and cob (McArdle 2011). Cobwork (tabya) was used in the Maghreb and Al-Andalus in the 11th and 12th centuries and was described in detail by Ibn Khaldun in the fourteenth century (1996: 766).

Today cob is still used in England, but it is widespread all over the world. One of the best examples of this technique in a modern building is the Methi School designed by Anna Heringer and Eike Roswag.

This project, carried out in Bangladesh, uses a cob wall to construct a living building. Anna Heringer has inserted some poles inside the wall which are used to keep the structure superimposed. Above the basement, it creates a secondary level, completely made with a bamboo structure. In this fine example, the cob wall represents the solid attachment to the ground of the building, a well-insulated space. So, using the plasticity of this technique, the architects created a very informal cave inside one of the main walls transforming the masonry into a playful and exclusive space for children.

>>>>>>>>>>>>>>>>>>

Torchis

Torchis is a constructive system that uses two different materials: earth and wood.

Wood is used for the construction of a load-bearing framework (usually a frame or half-timbered structure), while the worked earth mixture forms the filling. It can be used both for load-bearing walls and internal partitions.

The mixture of soil can be variable in size and consists of water, clay and natural fibres (straw, hay, vegetable fibres in general). In some geographical areas, it is even enriched with sand and lime, the latter to make it a waterproof aggregate. The consistency must be plastic (semi-fluid), such as natural concrete, with limited resistant function, so that it can be manually pressed into the timber frame.

In Africa, especially in Ethiopia, the mixture consists of earth-mud, and to make it termite soil is often used because it ensures greater resistance to compression, also natural fibres allow you to create a structural system with good mechanical resistance.

The wooden structure guarantees a good elasticity and deformability even to horizontal actions and its solidity is guaranteed by the mass of dried earth. To improve the resistance of the earth mixture a canopy or wooden slats

Fig. 04.07
Ethiopian torchis hut

Fig. 04.08
Torchis updated
student's maquette [Bonsembiante, Giordano, Vago]
"Necessary Architecture Workshop". Politecnico di Milano, February 2019

are often used between the main and secondary framework of the wooden structure. In this way, the grip of hand-pressed earth mixture in the interstices of the wooden frame is made easier.

Finally, the surface of the walls can be plastered with lime mortar.

Torchis is considered to be the first composite system in history because it is composed by a part of compression-resistant plastic material (earth or mud, clay), water and tensile-resistant materials (vegetable fibres or straw).

The racks were made of flexible and not too putrescible wood (hazelnut or split chestnut, for example), and assembled in the form of trays on the model of a simple basket. Thinner fibres, softened by soaking or cooking, were used for finishing between the top of the walls and the roof.

Torchis has been a building material used for European homes since the Neolithic period. The Celts used it a lot to build their walls. It is typical of the regions where clay soil (which cracks and sticks easily) has to be reinforced with straw (Houben, Guillaud 1998). This technique developed a lot from the fourteenth to the sixteenth century and lasted until the aftermath of the Second World War with a simple half-timbered wood. Today, torchis is mainly used for renovations in the western world but is still widespread in many southern countries.

The advantages of the material are that it provides thermal and acoustic insulation, is low cost, allows self-construction, is of local origin and has a very low impact on the environment (sustainable development), is a healthy material, there is no waste at the end of its life.

At the same time, the weakness of this type of technique is linked to its composition: the need to have a wooden structure binds this type of building to the possibility of obtaining wood. And the second weak point is represented by the size of the wall. Dimensions are normally defined by the size of the main wooden frame, which means that the final thickness of this type of wall will never be so great as to obtain the same performance as the other raw earth solutions.

In Africa, torchis spreads in countries rich in forests and alluvial soils (especially in the State of Ivory Coast), where man is a gatherer because agriculture is rarely possible, and buildings are temporary and quick to build.

In the walls of this technique, the earth is not the structural element, so it is easy to construct multi-level buildings because this is linked to the wooden structure.

Finally, this kind of wall needs to be covered with a plastering layer that protects both the mixture of earth and fibre and the timber structure.

Using torchis in a new building could be a very interesting way of dividing the two "natures" of this wall and applying them to different spatial situations. So, for example, if closed rooms could be built in the usual way (as described), the internal wooden structure could be used as a border boundary in open spaces. It is possible to define a unique system in which the earth wall changes its nature, allowing the wooden structure to highlight the presence of open space.

>>>>>>>>>>>>>>>

Earthbags

Earthbags buildings fill a unique niche in the quest for sustainable architecture.

Bags can be filled with local natural materials, which lowers the embodied energy commonly associated with the manufacture and transportation of building materials. Filling material is generally of mineral composition and is not subject to decomposition (even when damp), nor is it attractive to vermin or burnable, so it is extremely durable. It is also completely non-toxic and doesn't disperse harmful fumes into the building.

The walls are substantial, resist all kinds of severe weather and also stand up to natural calamities such as earthquakes and floods.

They can be erected simply and quickly with readily available components, for very little money.

Earthbags provide either thermal mass or insulation, depending on what they are filled with. When filled with soil, they provide thermal mass, when filled with lighter weight materials (such as crushed volcanic stone, perlite, vermiculite or rice hulls) they provide insulation.

On the other hand, when filled with gravel, bags can act as natural non-wicking, somewhat insulated foundations.

Because earthbags can be stacked in a large variety of shapes, including domes, they have the potential to virtually eliminate the need for common tensile materials in the structure, especially wood and steel which are often used for roofs.

This method is also very cheap, fill material can be on-site soil and the earthbags themselves can often be purchased as misprints or recycled grain sacks, but even when new they are not very expensive.

Polypropylene bags have good strength and durability, as long as they are kept away from too much sunlight.

For permanent buildings, bags should be covered with some kind of earthen plaster to protect them.

The earthbag solution was used for the first time by Nader Kalili, an Iranian architect employed by California University. He developed his Super Adobe system in 1984 (Kalili 1996), in response to a NASA call for designs for human settlements on the Moon and Mars. The project had been completely theoretical until the Persian Gulf War when refugees were sent into Iran. When this occurred Khalili partnered with the United Nations Development Programme (UNDP) and the United Nations High Commissioner for Refugees (UNHCR) and applied his research to emergency shelters. In 1991 he founded the California Institute of Earth Art and Architecture (Cal-Earth), where he taught his Superadobe building technique.

The Superadobe building shows the potential and the weakness of this earth solution: the potential is of course related to the possibility of realizing a building with one unique material and with the use of a very small quantity of water, two conditions that are very important in emergency situations. The weakness is related to the form: this kind of building is closely related to the igloo shape, that is a self-sustaining shape. By changing the shape, it is very important to integrate this technique with some other solutions for the covering system. It is also important to realize the weakness of this type of wall when it is used linearly, which is why it is very important to use, for example, buttresses to support the wall.

Openings

Raw earth remains one of the most widespread materials in the world. The "vernacular" origins of its application seem outdated, at least conceptually. However, it is still not possible to say that it is a material that has become part of the "ordinary" market.

On the other hand, even in non-emergency contexts, construction systems are beginning to look for an innovative solution using raw earth. From three-dimensional printing to certified bricks, from plasters to insulating systems to walls.

Unfortunately, there remains a profound detachment between emergency and ordinary contexts. In emergency contexts, earth is almost always a choice of necessity.

A careful choice, of course, which implies not only the possibility of creating efficient buildings but also of involving people in the construction process. Raw earth is a material that is well suited to the principles of self-construction.

This system, which involves the involvement of the "non-expert" population in the construction process of the work, adds a further qualification to this technique: the possibility of being easily transmitted. Following the Papert (1993) teaching principle, according to which errors are tools for improvement (Capponi 2008), the act of do-it-yourself becomes an act of learning

If, on the one hand, we have seen how world climate and economic changes directly affect vernacular architecture, understood in its primordial constitutive act (definition of materials, construction choices,

Fig. 04.09
Superadobe or Ecodome realized by Nader Kalili

Fig. 04.10
Earthbag used in line with student's still maquette [Matosevic, Seregni, Sheikhhassani, Ahsani] "Necessary Architecture Workshop". Politecnico di Milano, February 2019

shape), it is perhaps a good idea to end this chapter with a final reflection.

Climate and economic changes are widespread all over the world, and although their effects are less evident in consolidated contexts such as those in which we live, they remain an evident fact. The reflections on the widespread precariousness of society (Baumann 2001)[4] and the need of the people most affected by this precariousness to find new points of reference not only at work (Sennett 2008)[5] are known. Studies on new urban and non-urban figures who coexist in this socio-economic system are also known, such as the "working poor" of the metropolises (Petrillo 2006) forced to live below the poverty line in urban contexts (Martinotti 1993).

Finally, it is important to keep in mind both the constant migratory flow that affects the so-called rich countries, which generates a mass of poorly housed and often poorly managed people, and also the internal migratory flow of the younger population looking in abandoned villages and the countryside for an adequate housing solution both in terms of quality and economy.

This flow must be applied to the topic of fragile territories (Di Venosa, Morrica 2018) linked to the emergency of landscape protection which has highlighted the presence and the need to protect a large heritage built-in natural context that needs special care and attention (Pignatti et al. 2018).

This transition scenery must be assessed also considering how the recent health crisis has affected the "information society" in front of forced isolation and the economic crisis. This situation has emphasized the limits and potential of the smart society, or Society 5.0 (Önday 2019).[6] So, it is interesting to speculate how the growth reflections hypothesized for emergency contexts can be translated into non-emergency contexts.

Raw earth, like other solutions born from the vernacular and then hybridized, could become an effective alternative for new buildings and above all redevelopment of unconsolidated housing situations. The availability, cheapness, malleability and adaptive capacity of this material and these techniques make it highly topical in the face of peri-emergency contexts such as the one described above.

The relevance of these issues is also demonstrated by the interest of many architectural students on these topics. An increased number of specialized courses and workshops related to these themes is registered worldwide.

Furthermore, these materials and this type of technology (as described up to now) can be easily applied according to self-construction processes. Self-construction represents a principle of self-determination through which, in the construction act, multiple values are synthesized: the sense of belonging, the attachment, the cognitive ability of the technique. As Turner (1978) succinctly puts it, the self-construction process has more value than the intrinsic quality of the house, since it ties the house to its inhabitant. The basic inexperience of those who begin this type of process supports the link between traditional culture and contemporary culture on the basis of Levi Strauss' *La Pensée sauvage* (1964).

The "leap into the void" that serves to give shape to a work (Dieste 1996), taking into account the cognitive potential and the sense of appropriation, becomes a sure leap, as the quality of the result is shifted from the formal judgment to that of the participatory process.

It can be affirmed that the teaching of the vernacular, not only specific but conceptual, linked to knowing how to adapt to build not only the minimum necessary but a comfortable living environment, not only is still very topical but could in the future be a solution in unexpected contexts.

>>> Notes

1 "South Africa: Poverty reduction winners". IRIN. 11 January 2002. www.development-marketplace.org.

2 "Housing and Jobs for a Better Future". It is a project of the Development Marketplace program of the World Bank. [https://web.archive.org/web/20131113032657/http://wbi.worldbank.org/wbdm/]. The Development Marketplace (DM) is a competitive grant programme that identifies and funds innovative, early stage development projects that are scalable and/or replicable, while also having high potential for development impact ... Since its inception in 1998, the DM programme – housed in the World Bank and administered by the World Bank Institute – has awarded more than $60 million in grants to more than 1,200 innovative projects. All of these projects have been identified through country, regional and global DM competitions. Using DM funding as a launch pad, many projects have gone on to secure additional funding support from other donors, foundations, governments, impact and CSR Investors.

3 "Cobbing" [verb]: (transitive) British informal to "beat", esp on the buttocks Collins English Dictionary (HarperCollins).

4 Baumann gives this type of society the character of "liquidity". The term aims to give a sense of temporariness and precariousness (Baumann 1999) which characterized most of the working and living situations. These conditions are widespread in all working conditions and can sometimes lead to those needed for the community, in the sense of relationships and ties with people, which are lacking in today's society (Baumann 2002).

5 According to the theory of Richard Sennett, the reason for the dissatisfaction and unhappiness of human beings in modern society is the weakness of temporary links with other people (Sennett 1998). One of the answers that humanity offers to these weak conditions is to find strength in unofficial lifetime: the lifetime of the hobby, rather than working hours, has become an opportunity to find satisfaction and happiness (Sennett 2008).

6 The term refers to the idea that Society 5.0 will follow Society 1.0 (hunter-gatherer), Society 2.0 (agricultural), Society 3.0 (industrialized) and Society 4.0 (information). Also called the "super-smart society, Society 5.0 envisions a sustainable, inclusive socio-economic system, powered by digital technologies such as big data analytics, artificial intelligence (AI), the Internet of Things and robotics. The "cyberphysical system", in which cyberspace and the physical space are tightly integrated becomes a pervasive technological mode supporting Society 5.0. To some, the concept sounded like a lofty vision at first, with no clear prospect of how it would actually unfold (en.unesco.org).

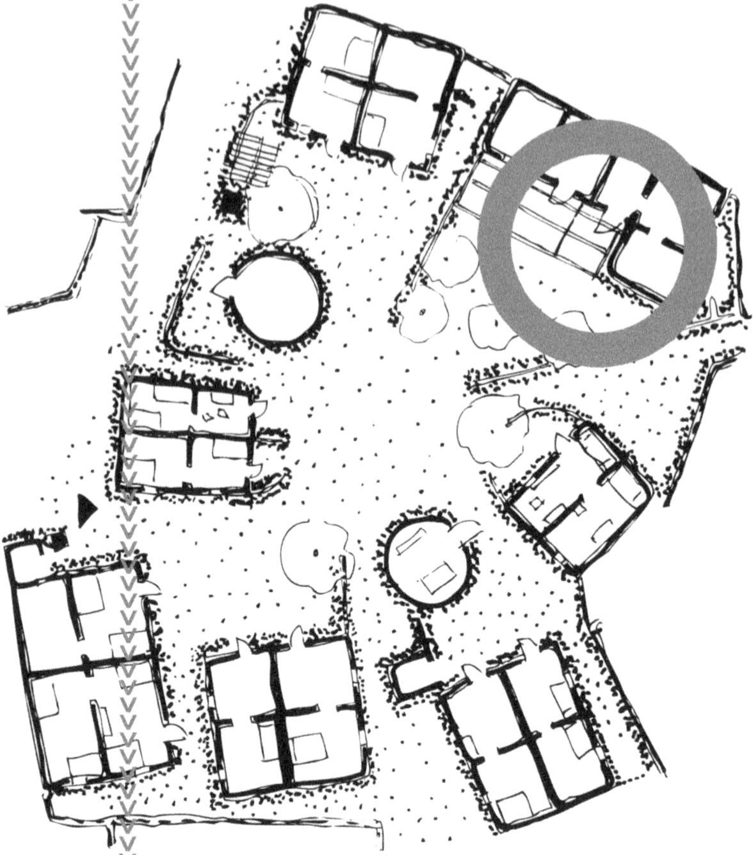

To the future for ***fragile*** *but* ***strong*** *territories*

05

Alisia Tognon | Luca Trabattoni

DOI: 10.1201/9781003157632-05

The study efforts discussed in the previous chapters aimed to understand the topological, morphological and typological theoretical model in conjunction with the continual emphasis placed on the Nigerien context and were, therefore, site-specific. Moreover, we believe that the relevance of architecture does not lie in the understanding of technical details, but the recognition that architecture is inevitably a social act and the result of collective memory. The environment reflects a set of culture, whereby the local landscape becomes the ground zero for understanding the answers of human populations around the world. Human interactions, personal beliefs, traditional and evolving technologies, situational opportunities and constraints, create, adapt and transform the environments in response to the forces of the natural environments. Reading human inscriptions on earth allows an understanding of the social factors in each region and how the built environment has evolved through human intervention over time.

Working on Niger's landscape, there are probably more opportunities than challenges. The particularly arduous climatic conditions, linked with critical socio-political conditions and widespread scarcity of resources, both economic and material, make life in Niger potentially demanding. However, in this context, it is possible to figure out the way to work and adapt to the place, understanding what we have at our disposal.

Undoubtedly, architecture is not the overall answer to this widespread condition. However, a better quality of life is also achieved through the correct design of living spaces in the houses.

Niger, like other African countries, brings with it an extraordinarily dense and fascinating cultural background, which will inevitably be transliterated in the definition of its architecture. Over the years, the ability to interpret the climate and the geographical environment has resolved the construction of high-efficiency dwellings, both for the use of local resources and for habitat performance, based on passive and low–impact principles. In line with this, we do not have not to break with the past, but we should push it forward, adding new technology to it.

In light of this, raw earth is the main protagonist of this knowledge, that is getting lost. As we analysed in this book, for centuries it has been the primary material for construction in Niger, due to the high local material retrieval, its extremely efficient quality and adaptability to numerous solutions. Moreover, technology is effortless and elementary for self-building. As we have already debated raw earth has overcome the vernacular shape limits by adapting to the contemporary architecture languages, pursuing at the same time both efficiency and a strong identity with the local building traditions.

Adhering to the well-known notions of genius loci and tectonic culture (Norberg Schulz 1982; Frampton 1995), the design and technic solutions of these dwellings is a process that explores the perennial qualities of architecture through its archetypes of floor/wall/roof and inside/outside. In the Nigerien landscape, the character and sense of place become evident in the way boundary of the enclosure (fences and also floors, walls and ceilings) is articulated or constructed within the space.

However, one of the traumatic components in Niger is represented by climate change and also by the socio-political intrusions of the western economy (D'Onofrio 2015), which in addition to having undermined Niger's development for years by exploiting its natural resources, inevitably introduces a technological leap that influences the Nigeriens to adapt. In the construction sector, the most common reaction is the imitative one, which aims at the use of "new" materials from western technics such as concrete and sheet metal, not considering its unsuitability in the face of local climatic conditions. Conversely, it is believed that reflection on building needs in these contexts implies an awareness of the resilient potential of vernacular architecture. The cultural heritage determines the reactive bases for proposing practical solutions.

Faced with the issue of emergency and sustainability, one of the most incisive contemporary responses concerns the issue of resilience. The term "resilience" has become relevant in the architectural field and derives from the Latin *re-* "back" and *salīre* "jump", or "go back" (Garzanti Linguistica 2015), and indicates the ability to endure traumas, breakages and events, and then return to its original state (The Resilience Alliance 2002). In essence, resilience is the ability of an ecosystem to react and adapt to external agents, tolerating disturbance without collapsing into a qualitatively different state that is controlled by a different set of processes. For example, in biology, a resilient object or substance can recover its shape after trauma. Similarly, in ecology, we define a resilient ecosystem as one that is able to resist any disruption, without permanent damage and with the capability to return to the initial conditions. In the 1950s the term "resilience" was used in psychology to describe the tolerance abilities of children, as well to identify a series of qualities that allow people to remain psychologically balanced and mentally healthy in the case of adverse life circumstances and crisis (Petzold, Muller 2004).

In recent years the term "resilience" has gained relevance in various disciplines and scientific contexts (Burkner 2010). With regard to regional architecture, its principles derive from the "climate-adapted" design over time, social behaviour and cultural identity. In light of this, the concept of resilience appears to be bound to the approaches taken by human ecology and taxonomy in studies on developing countries, in that a resilient ecosystem can withstand shocks and rebuild itself when necessary.

Making a comparison, therefore, between vernacular architecture and the parameters that define a resilient system allows us to identify some substantial common arguments (Jha, et al. 2013; Langeveld 2013).

The simplicity of vernacular architecture in terms of construction technique allows a more manageable level of maintenance for energy and costs and can be adapted to changing conditions.

The concept of resilience also includes the ability to use local resources, both for energy and building materials. Traditional architecture always worked with a "passive climate system",[1] which lets to be allows a self-sustainable system independent from the non-renewable resources.

The vernacular paradigms could represent a redundant system. As the use of raw earth has demonstrated, concerning the context, a similar approach brings to different solutions. The unending variations answer to several users with the same quality and form on their own/independently to respond to shocks and changes, in terms of communities, ecosystems, economies and social systems. In line with this is the increased durability of the system and, consequently, the life cycle of buildings, which includes building practices, design, infrastructure and ecosystems.

Another crucial point is the so-called "know-how", which makes the communities acquire greater resilience. The transmission of "know-how" increases the capacity of "recoverability" in case of perturbations. The resilience is noteworthy for the mitigation of the damage caused by natural disasters. In that sense, the transmission of habitat cultures between generations increases the acknowledgement of basic needs and defines simple and well-known construction cultures by the inhabitants. In this way, the self-building approach related to the knowledge of vernacular tradition increases the "know-how" dissemination of living cultures and increases the adaptability of the building answers.

A state of necessity means a lack of physical and cultural resources. In this situation, the emergency defines a poor and intricate context. The potential development and improvement of this circumstance involve a knowledge of the historical tangible and intangible heritage, with an in-depth analysis of the contemporary situation.

In that sense, vernacular architecture is an integral part of the societies and cultures that produced it. It becomes a relevant element in our research because it shows up the different and distinctive ways, which are also elegant and ingenious, through which populations, all over the world and over time, have imagined, designed, used and maintained their built environments. Economic models, cultural values, political relationships, religious beliefs and social structures all have an intricate relationship with architecture, determining which materials are used, which shapes are chosen, how space is used and they result in the enormous diversity of construction methods that characterize the world's vernacular architecture.

A holistic and integrated perspective that examines all aspects of a building tradition (social, economic, political and environmental) and how they connect is essential. We believe that the main lesson that vernacular architecture can teach is related to the value and importance of cultural diversity, and also the ability to find a suitable solution to local emergency. The use of raw earth is only one of the aspects that could be learned. The implementation of this system, hybridized with contemporary material, could be an adequate solution to the real need.

Alongside the technical considerations, it is fundamental to also support the need for community participation, without which the previously mentioned aspects have a short future.

The notion of "sustainability" in such contexts has value if applied to the concept of "support for people". In African countries, sustainability is not something that can be measured in terms of certification. On the contrary, supporting people simultaneously helps in to support the environment. The use of local materials provides economic advantage, educational provision for the continuation of traditions and environmental support. Following this idea, we are able to create a positive and inclusive environment from a social point of view: having developed an extraordinary attachment to the buildings, people become more environmentally friendly.

The method of participation is not easy to do with ordinary tools but has to be established as the building is being constructed. Self-construction, therefore, applied to raw earth and the rediscovery of the vernacular tradition is a field to be taught and translated into modern times.

An approach to raw earth is therefore necessary, starting from the principles of self-construction as a system of transmission of knowledge, from experts to non-experts, with the intention, even before creating great architecture, of training new specialists. We believe in sustainability in terms of "people": supporting people as it simultaneously helps us to support the environment.

Even if an interest in updating traditional techniques through technology in Niger has not yet begun, we noted some curiosity about the idea, and some projects are taking that direction.

>>> **Note**

1 Thermal mass, cross ventilation, orientation and the other passive way to approach sustainability are at the base of the vernacular architecture.

Bibliography

>>>Ch [00]

Espuelas, F. (2009), *Madre Materia*, Milano: Christian Marinotti Editore.

Fathy, H. (1969), "Tradition's role", in *Architecture for the Poor: An Experiment in Rural Egypt, Chicago and London:* The University of Chicago Press, pp. 24–27.

Rudofsky, B. (1964), *Architecture Without Architects: A Short Introduction to Non-Pedigreed Architecture*, New York: The Museum of Modern Art.

Schumacher, F. (1973), *Small is Beautiful: A Study of Economics as if People Mattered*, London: Blond & Briggs.

>>>Ch [01]

Abbado, M., Adamou, D., Aliou, I., Assoumane, B., Kalilou, O. (2015), *L'architecture en Terre au Niger Entre Passe et Future*, Niamey: La Nouvelle Imprimeire du Niger.

Agency for Technical Cooperation and Development (ACTED) (2012), *Flooding in Niger: Stopping History Repeating Itself*. Available online: http://www.acted.org/en/flooding-niger-stopping-history-repeating-itself [access on 7th April 2020].

Aker, J. C., Klein, M. W., O'Connell, S. A. (2016), "International and Intranational Market Segmentation and Integration in West Africa", in *African Successes, Vol. IV: Sustainable Growth*, Edwards, S., Johnson, S., Weil, D. N. (eds.), Chicago and London: The University of Chicago Press, pp. 179–199.

Alesina, A., Baquir, R., Easterly, W. (1997), *"Public Goods and Ethnic Divisions", Quarterly Journal of Economics* 114 n. 4, pp. 1243–1284. Available online: https://dash.harvard.edu/handle/1/4551797 [access on 23rd April 2020].

Allen, F. (2012), *Implementation of Oil Related Environmental Policies in Nigeria: Government Inertia and Conflict in the Niger Delta*, Newcastle: Cambridge Scholars Publishing.

Aloko, N. J., Diallo, A., Motcho, K. H. (2010), *Towns and Organization of Space in Africa*, Paris: Karthala Editions.

Amani, A., Dessouassi, R., Paintsil, A. (2015), *"Integrated water resources management: African perspectives"*, in *Sustainability of Integrated Water Resources Management: Water Governance, Climate and Ecohydrology*, Setegn, S. G., Donoso, M.C. (eds), Cham (CH): Springer, pp. 25–50.

Ambouta, K. (1984), *Contribution a L'édaphologie de la Brousse Tigrée de L'ouest Nigérien*, Ph.D. Thesis in Engineering, Université de Nancy I.

Arecchi, A. (1991), *La Casa Africana*, Milan: CLESAV-Città studi.

Arecchi, A., Delisse, L. F. (1999), *Architettura Magica. le Facciate Ricamate di Zinder, Capitale Degli Haussa del Niger*. Milan: Mimesis Edizioni.

Baudouï, R., Potie, P. (2003), *André Ravéreau L'atelier du Désert*, Marseille: Editeur Parenthèses.

Bernacsek, G. M., Hughes, J. S., Hughes, R. H. (1992), *A Directory of African Wetlands*. Gland: IUCN, Nairobi: UNEP, Cambridge: WCMC.

Bernus, E., Laclavère, G., Sidikou, A. H. (1980), *Atlas du Niger*. Paris: Editions Jeune Afrique.
Boudet, G. (1969), "*Projet de mise en valeur du Dallol-Maouri (rép. du Niger): étude des pâturages naturels*", *Études Agrostologiques*, n. 26, Maisons Alfort, IEMVT. Available online: https://docplayer.fr/93857667-Etude-des-paturages-naturels.html [access on 22nd March 2020].
Bourdier, J. P., Minh-ha, T. (2005), *Habiter un Monde. Architectures de l'Afrique de l'Ouest*, Paris: Editions Alternatives.
Brachet, J. (2012) "*Movements of people and goods: local impacts and dynamics of migration to and through the Central Sahara*", in *Saharan Frontiers: Space and Mobility in Northwest Africa*, Scheele, J., McDougall, J. (eds.), Bloomington: Indiana University Press, pp. 238–256.
Chatwin, B. (1987), *The Songlines*, London: Jonatan Cape; trad. it. 1995, Le vie canti, Milan: Adelphi.
CILSS (Comité Permanent Inter-Etats de Lutte contre la Sécheresse dans le Sahel) (2016), *Landscapes of West Africa: A Window on a Changing World. U.S. Geological Survey EROS*, 47914 252nd St, Garretson, SD 57030, United States. Available online: https://eros.usgs.gov/westafrica/sites/default/files/ebook-English/index.html#p=4 [access on 17th March 2020].
Clements, P. C. (1905), *Research Methods in Ecology*. Lincoln Nebraska: University Publishing Co.
Collier, P. (2007), *The Bottom Billion*, New York: Oxford University Press.
Décamps, H., Fortuné, M., Gazelle, F., Pautou, G. (1988), "*Historical influence of man on the riparian dynamics of a fluvial landscape*", *Landscape Ecology* 1, pp. 163–173.
Di Castri, F., Hansen, A. J., Holland, M. M. (1988), "*A new look at ecotones: emerging international projects on landscape boundaries*", *IUBS, Biology International*, Special Issue AJH 17. Available online: http://www.iubs.org/fileadmin/user_upload/Biology-International/BI-Specials/SPECIAL_ISSUE_17a.pdf [access on 20th March 2020].
Easterly, W., Levine, R. (1997), "Africa's growth tragedy: policies and ethnic divisions", *The Quarterly Journal of Economics*, Vol. 112, n. 4, pp. 1203–1250.
FAO (Food and Agriculture Organization) (2016), Country data collection of Niger. Available online: http://www.fao.org/countryprofiles/index/en/?iso3=NER [accessed on 15th April 2020].
Farina, A. (1995), *Ecotoni, Patterns e processi ai margini*, Padova: CLEUP.
Farina, A. (1998), *Principles and Methods in Landscape Ecology*. Cham (CH): Springer-Science+Business Media, B.V.
FAOSTAT Food and Agriculture Organization of the UN (2011). Available online: http://faostat.fao.org/site/550/default.aspx#ancor [access on 14th March 2020].
Fraser, B. G., Williams, D. D. (1998), "*Seasonal boundary dynamics of a groundwater/surface-water ecotone*", *Ecology*, 79, n. 6, pp. 2019–2031.
Gallais, J. (1984), *Hommes du Sahel: Espaces-Temps et Pouvoirs. Le Delta Intérieur du Niger, 1960–1980*, Paris: Flammarion Edition.
Galle, S., Ehrmann, M., Peugeot, C. (1999), "*Water balance in a banded vegetation pattern: a case study of tiger bush in western Niger*", *Catena*, 37, n. 1–2, pp.

197–216. Available online: https://www.sciencedirect.com/science/article/abs/pii/S0341816298900601 [access on 17th April 2020].
Gavaud, M. (1966), *Etude Pédologique du Niger Occidental*, Dakar: Centre ORSTOM de Hann.
Hansen, A. J., Di Castri, F., Nauman, R. J. (1992), "Ecotones: what and why?", in *Landscapes Boundaries: Consequences for Biotic Diversity and Ecological Flows*, Hansen, A. J., Di Castri, F. (eds.), New York: Springer Verlag, pp. 9–46.
Harris, L. D. (1988), "Edge effects and conservation of biotic diversity", *Conservation Biology*, n. 2, pp. 330–332.
Hockings, P. (1995), *Principles of Visual Anthropology*, Berlin & New York: Mouton de Gruyter.
Houben, H., Guillaud, H. (2006), *Traité de Construction en Terre, CRATerre, L'ency-Clopédie de la Construction en Terre, Vol. I.*, Marseille: Editions Parenthèses.
Human Development Report (2019), *Inequalities in Human Development in the 21st Century*. Avalable online: http://hdr.undp.org/sites/all/themes/hdr_theme/country-notes/NER.pdf [access on 16th March 2020].
Idrissa, A., Decalo, S. (1992), *Historical Dictionary of Niger*, Lanham Maryland: Scarecrow Press.

INS (Institut National de la Statistique) (2019), Publications Démographiques et Sociales. Available online: https://www.stat-niger.org [accessed on 15th April 2020].
Kadafa, A. A. (2012), "Environmental impacts of oil exploration and exploitation in the Niger Delta of Nigeria", *Global Journal of Science Frontier Research Environment & Earth Sciences*, 12, n. 3, Online Global Journals Inc. (USA).
Kultermann, U. (1963), *Architecture Nouvelle en Afrique*, Paris: Edition Albert Morancé.
Le Houerou, H. N. (1989), The Grazing Land Ecosystems of the African Sahel, Berlin: Springer.
Li, H., Reynolds, J. F. (1994), "A simulation experiment to quantify spatial heterogeneity in categorical maps", *Ecology, 75*, pp. 2446–2455.
Loflin, C. (1998), *African Horizons: Landscapes of African Fiction*, Westport Connecticut: Greenwood Press.
Lowman, P. D. (1968), *Space Panorama*, Zurich: M. A. Müller.
Lynch, P. A., Roberts, J. (2004), *African Mythology*, A to Z, New York: Chelsea House.
Maignien, R. (1966), *Review of Research on Laterite, Natural Resource Research IV*, Paris: UNESCO.
Mauro, P. (1995), "Corruption and growth", *The Quarterly Journal of Economics*, 110, n. 3, Cambridge Massachusetts: The MIT Press, pp. 681–712.
Merriam-Webster (2021), Ecotone, in *Webster's Dictionary and Thesaurus of the English Language*, Springfield, Massachusetts. Available online: https://www.merriam-webster.com/dictionary/ecotone [accessed on 15th April 2020].
Mester de Parajd C. and L. (1988), *Regards sur L'habitat Traditionnel au Niger. Les Cahiers de Construction Traditionnelle*, Nonette: Édition Créer.
Moestrup, S. (1999), "The role of actors and institutions: the difficulties of democratic survival in Mali and Niger", *Democratization*, n. 6, pp. 171–186.

Morris, J. (2005), *Butabu: adobe architecture of West Africa*, Hudson: Princeton Architectural Press.

Mukerji, K., Roland, S. (1981), *Appropriate Building Materials: A Catalogue of Potential Solutions*. Bradford: ITDG Publishing.

OCHA (2012), *Snapshot of Floods*. Available online: https://reliefweb.int/sites/reliefweb.int/files/resources/Snapshot_Inondations%2017%2009%202013.pdf [access on 7th April 2020].

Okoh, R. N. (2006), "Conflict management in the Niger Delta Region of Nigeria: a participatory approach in the poverty of the Niger Delta Nigerian economy society", in *Fiscal Federation and Nigerian Economy Development*, Annual Conference Abuja Nigeria.

Peltier, R., Bertrand, A., Lawali, M., Madon, G., Montagne, P. (1997), "Marchés ruraux de bois-énergie au Sahel", *Bois forêts tropiques* 245, n. 2, pp. 75–89. Available online: https://agritrop.cirad.fr/388733/1/document_388733.pdf [access on 15th April 2020].

Pickett, S. T. A., White, P. S. (1985), *The Ecology of Natural Disturbance and Patch Dynamics*, New York: Academic Press.

Prussin, L. (1974), "An introduction to indigenous African architecture", *Journal of the Society of Architectural Historians*, 33, n. 3, Berkeley: University of California Press on behalf of the Society of Architectural Historians, pp. 182–205.

Red Cross and Red Crescent (2010), "NIGER: Floods", DREF operation n° MDRNE006 11 August. Available online https://www.ifrc.org/docs/appeals/10/MDRNE006do.pdf [access on 7th April 2020].

Reitano, T., Shaw, M. (2015), "Fixing a fractured state? Breaking the cycles of crime, conflict and corruption in Mali and Sahel", *The Global Initiative Against Transnational Organized Crime*, VI, Geneva.

Rochette, R., Gronoff, J. D., Masseport, F. (1966), *Doumέga, Dioundiou, Kawara Débé villages des Dallols, Maouri et Fogha: Monographies Comparées*, Niamey: IFAN.

Salifou, A. (2002), *Le Niger*, Paris: Editions L'Harmattan.

Sauer, P. (1995), "Placemarks: ecotones", *Orion*, n. 14, pp. 59–61.

Schulz, E., Abichou, A., Adamou, A. (2009), "The desert in the Sahara. Transitions and boundaries", in *Holocene Palaeoenvironmental History of the Central Sahara*, Baumhauer, R., Runge, J. (eds.), *Palaeoecology of Africa, Vol. 29, An International Yearbook of Landscape Evolution and Palaeoenvironments*, London: Routledge.

Schutyser, S., Dethier, J., Eaton, R., Gruner, D. (2003), *Banco, Mosquées en Terre du Delta Intérieur du Fleuve Niger*, Milan: 5 Continents.

Spadari, L. (2017), *Morfologia Dinamica Dell'abitare. Caso Studio Città di Dosso*, Niger, Master's Thesis in Architecture and Engineering Construction, Politecnico di Milano.

Stoller, P. (2009), *The Power of the Between: An Anthropological Odyssey*, Chicago & London: The University of Chicago Press.

Taton, D. (1984), *Architecture traditionnelle au Niger, Vol. 32 of Human Settlements and Socio-Cultural Environment, Institute of Advanced Architectural Studies*, Paris: UNESCO.

The World Bank (2005), *Public Expenditure Management and Financial Accountability in Niger*, Washington D.C.: The International Bank for Reconstruction and Development.
Thorp, M. B. (1969), "Some aspects of the geomorphology of the Aïr Mountains, Southern Sahara", *Transactions of the Institute of British Geographers*, n. 47, pp. 25–46. Available online: JSTOR, doi:10.2307/621733 [access on 21st April 2020].
Tinti, P., Reitano, T. (2016), *Migrant, Refugee, Smuggler, Saviour*, London: Hurst.
UN Data (2019), *Niger*. Available online: http://data.un.org/en/iso/ne.html [access on 14th March 2019].
UNHCR (2019), *West Africa Mixed Movement Report*, July/September. Available online: https://data2.unhcr.org/en/documents/details/72013 [access on 14th March 2020].
USAID (2014), *Climate Change and Conflict in the Sahel. A Policy Brief on Findings from Niger and Burkina Faso*, January. Available online: https://www.climatelinks.org/sites/default/files/asset/document/Sahel_Case_Study.pdf [access on 14th March 2020].
Walraven, van K. (2003), "Sawaba's rébellion in Niger (1964–1965): narrative and meaning", in *Rethinking Resistance: Revolt and Violence in African History*, Abbink, J., de Bruijn, M., Walraven, van K. (eds.), Leiden: Brill Academic Publishers, pp. 218–252.
Wiafe Amoako, F. (2019), *Africa 2019–2020*, Lanham Maryland: Rowman & Littlefield Publishing Group.
World Population Review (2020), *Niger*. Available online: https://worldpopulationreview.com/countries/niger-population/ [access on 15th April 2020].

>>>Ch [02]

Abbado, M., Adamou, D., Aliou, I., Assoumane, B., Kalilou, O. (2015), *L'architecture en terre au Niger entre passé et future*, Niamey: La Nouvelle Imprimeire du Niger.
Allsopp, B. (1970), *The Study of Architectural History*, New York: Studio Vista.
Al-Qawasmi, J. A. (2014), "Contemporary vernacular as a diffusion of an innovation process", in *Vernacular Heritage and Earthen Architecture: Contributions for Sustainable Development*, Correia, M., Carlos, G., Rocha, S. (eds), London: Taylor & Francis Group, pp. 233–238.
Alexander, C. (1968), *Notes on the Synthesis of Form*, Cambridge Massachusetts: Harvard University Press.
AlSayyad, N. (2004), *The End of Tradition?*, London: Routledge.
Arecchi, A. (1991), *La casa africana*, Milan: Clesav Città Studi.
Arecchi, A. (1998), *Abitare in Africa: architetture, villaggi e città nell'Africa Subsahariana dal passato al presente*, Milan: Mimemis & Liutprand.
Arecchi, A., Delisse, L. F. (1999), *Architettura magica. Le facciate "ricamate" di Zinder, capitale degli Haussa del Niger*, Milano: Mimesis.
Asquith, L., Vellinga, M. (2006), *Vernacular Architecture in the Twenty-First Century: Theory, Education and Practice*, Abingdon: Taylor & Francis.

Bello, Y. O. (2001), "Elements of creativity in traditional architecture: A descriptive study of theory of architecture", *A Journal of AARCHE (Association of Architectural Educators in Nigeria)*, 1, n. 6, pp. 95–13.

Bidi, A. (2000), "La route du sel", in *Regional Thematic Expert Meeting on Cultural Landscapes in Africa; The World Heritage Convention and Cultural Landscapes in Africa*, Tiwi, Kenya, 9–14 March 1999, pp. 93–96.

Bourdier, J. P., Minh-ha, T.T. (2005), *Habiter un monde. Architectures de l'Afrique de l'Ouest*, Paris: Edition Alternatives.

Braungart, M., McDonough, W. (2002), *Cradle to Cradle: Remaking the Way we Make Things*, New York: Farrar, Straus and Giroux.

Bravmann, R. (1974), *Islam and Tribal Art in West Africa*, Cambridge: Cambridge University Press.

Carrol, K. (1992), *Architectures of Nigeria: Architectures of the Hausa and Yoruba Peoples and of the Many Peoples Between Tradition and Modernization*, London: Ethnographica Society of African Missions, Lester Crook Academic Publishing.

Correia, M., Carlos, G., Rocha, S. (2014), *Vernacular Heritage and Earthen Architecture: Contributions for Sustainable Development*, London: Taylor & Francis Group.

Cox, M. (2009), *Living in the New Millennium: Houses at the Start of the 21st Century*, London & New York: Phaidon.

Curtin, P. D. (1967), *Africa Remembered: Narratives by West Africans from the Era of The Slave Trade*, Madison: University of Wisconsin Press.

Daoudi, N. S., Mestoul, D., Lamraoui, S., Boussoualim, A., Adolphe, L., Bensalem, R. (2019), "Vernacular architecture in arid climates: adaptation to climate change", in *Bioclimatic Architecture in Warm Climates: A Guide for Best Practices in Africa, Correia Guedes*, M., Cantuaria, G. (eds.), Cham: Springer, pp. 119–154.

Denyer, S. (1978), *African Traditional Architecture*, London: Heinemann.

Echard, N. (1967), "L'habitat traditionnel dans l'Ader. Pays hausa, République du Niger", *L'Homme*, 7, n. 3, pp. 48–77. Available online: https://www.jstor.org/stable/40589771 [access on 14th April 2019].

Eglash, R. (1999), *African Fractals: Modern Computing and Indigenous Design*, New Brunswick NJ: Rutgers University Press.

Elleh, N. (1997), *African Architecture: Evolution and Transformation*, New York: McGraw-Hill.

Fassassi, M. A. (1978), *L'architecture en Afrique Noire, Cosmoarchitecture*, Paris: L'Harmattan.

Fathy, H. (1986), *Natural Energy and Vernacular Architecture: Principles and Examples with Reference to Hot Arid Climates*, Chicago and London: The University of Chicago Press.

Frampton, K. (1983), "Towards a critical regionalism: six points for an architecture of resistance", in *The Anti-Aesthetic. Essays on Postmodern Culture*, Foster, H. (ed.), Seattle: Bay Press, pp.16–30.

Frey, P., Bouchain, P. (2010), *Learning from Vernacular*, Arles: Actes Sud.

Friedman, A. V. (2012), *Fundamentals of Sustainable Dwellings*, Washington, DC: Island Press.

Fromonot, F. (2003), *Glenn Murcutt: Buildings+Projects 1962–2003*. London and New York: Thames & Hudson.
Gallais, J. (1984), *Hommes du Sahel: espaces-temps et pouvoirs. Le delta intérieur du Niger, 1960–1980*, Paris: Flammarion Edition.
Gardi, R. (1973), *Indigenous African Architecture*, New York: Van Nostrand Reinhold.
Garlake, P. (1978), *The Kingdoms of Africa: The Making of the Past*, Oxford: Elsevier-Phaidon.
Glancey, J. (2003), *The Story of Architecture*, London: Dorling Kindersley.
Gould, J., Kolb, W. L. (1964), *A Dictionary of the Social Sciences*, New York: The Free Press of Glencoe.
Guidoni, E. (1974), "Antropomorfismo e zoomorfismo nell'architettura primitiva", in *L'Architettura Cronache e Storia*, n. 222, pp. 752–763.
Guidoni, E. (1975), *Architettura primitiva*, Milan: Electa.
Heath, K. W. (2003), "Defining the nature of vernacular", *Material Culture*, 35, n. 2, pp. 48–54.
Hillier, B., Hanson, J. (1984), *The Social Logic of Space*. Cambridge: Cambridge University Press.
Houben, H., Guillaud, H. (2006), *Traitè de construction en Terre, CRATerre*, Marseille: Editions Parenthéses. Available online: https://unesdoc.unesco.org/in/documentViewer.xhtml?v=2.1.196&i...=en&multi=true&ark=/ark:/48223/pf0000025219/PDF/025219freb.pdf [access on 7th April 2019].
Idrissa, A. (2020), *Historical Dictionary of Niger*, London: Rowman & Littlefield.
James-Chakraborty, K. (2014), "Africa: villages and cities", in *Architecture Since 1400*, Minneapolis: University of Minnesota Press, pp. 411–423.
Jencks, C., Baird, G. (1969), *Meaning in Architecture*, London: Barrie and Rockcliff.
Larousse (2021), Mihrab. Available online: https://www.larousse.fr/dictionnaires/francais/mihrab/51402 [accessed on 15th April 2020].
Lebeuf, J. P. (1967), "L'architecture africaine traditionelle", *Colloque sur l'Art nègre du 1er Festival Mondial des Arts Nègres*, Paris: Présence Africaine, pp. 319–328.
Lynch, P. A., Roberts, J. (2004), *African Mythology, A to Z*, New York: Chelsea House.
Maudlin, D. (2010), "Crossing boundaries: revisiting the thresholds of vernacular architecture", *Vernacular Architecture*, 41, pp. 10–14. Available online: https://www.tandfonline.com/doi/pdf/10.1179/174962910X12838716153682 [access on 16th April 2019].
McDonough, W. (1996), "Design, ecology, ethics, and the making of things", in *Theorizing a New Agenda for Architecture*, Nesbitt, K. (ed.), Princeton: Princeton Architectural Press, pp. 400–407.
Mester de Parajd C. and L. (1988), *Regards sur l'habitat traditionnel au Niger. Les cahiers de Construction Traditionnelle*, Nonette: Édition Créer.
Mileto, C., Vegas, F., García Soriano, L., Cristini, V. (eds.) (2015), *Vernacular Architecture: Towards a Sustainable Future*, Proceedings of the International Conference on Vernacular Heritage, Sustainability and Earthen Architecture, Valencia, Spain (11th–13th September 2014), London: Taylor & Francis Group.

Mumford, L. (1961), *The City in History: Its Origins, its Transformations, and its Prospects*, San Diego: Harcourt, Brace & World.
Norberg Schulz, C. (1971), *Existence, Space and Architecture*, New York: Praeger Publishers.
Odeyale, O., Adekunle, T. O. (2008), "Innovative and sustainable local material in traditional African architecture: socio cultural dimension", in *Structural Analysis of Historic Construction: Preserving Safety and Significance*, D'Ayala, D., Fodde, E. (eds.), London: Taylor & Francis Group. Available online: DOI: 10.1201/9781439828229.ch113 [access on 16th April 2019].
Oliver, P. (1969), *Shelter and Society*, London: Barrie and Rockliff.
Oliver, P. (1971), *Shelter in Africa*, London: Praeger Publishers.
Oliver, P. (1984), "Round the houses", in *British Architecture*, Papadakis, A. (ed.), London: Architectural Design.
Oliver, P. (1987), *Dwellings: The House Across the World*, Austin: University of Texas Press.
Oliver, P. (1997), *Encyclopedia of Vernacular Architecture of the World*, Cambridge: Cambridge University Press.
Oliver, P. (2003), *Dwellings: The Vernacular House World Wide*. London: Phaidon.
Oliver, P. (2006), *Built to Meet Needs: Cultural Issues in Vernacular Architecture*, Oxford: Elsevier.

Preston Blier, S. (1987), *The Anatomy of Architecture: Ontology and Metaphor in Batammaliba Architectural Expression*, London & Chicago: The University of Chicago Press.
Prussin, L. (1974), "An introduction to indigenous African architecture", in *Journal of the Society of Architectural Historians*, 33, n. 3, Berkeley: University of California Press on behalf of the Society of Architectural Historians, pp. 182–205.
Prussin, L., Adan, A. (1995), *African Nomadic Architecture: Space, Place, and Gender*, Washington: Smithsonian Institution Press.
Ramir, S. (1991), *Les pistes de l'oubli. Touaregs au Niger*, Paris: Editions du Félin.
Rapoport, A. (1969), *House, Form and Culture, Englewood Cliffs*, NJ: Prentice-Hall.
Rapoport, A. (2005), *Culture, Architecture, and Design*, Chicago: Locke Science Pub.Co.
Rasmussen, S. (1996), "The tent as cultural symbol and field site: social and symbolic space, topos, and authority in a Tuareg Community", *Anthropological Quarterly*, 69, n. 1, pp. 14–26.
Reader, H. J. (1999), *Africa: A Biography of the Continent*. New York: Knopf.
Redfield, R. (1965), *The Primitive World and Its Transformations*, Ithaca, New York: Cornell University Press.
Ricoeur, P. (1965), "Universal civilization and national cultures", in *History and Truth*, Evanston, IL: Northwestern University Press.
Roberts, W.E. (1985), "Untitled essay", *Material Culture*, 17, pp. 89–93.
Rudofsky, B. (1964), *Architecture Without Architects: An Introduction to Non-Pedigreed Architecture*, New York: MOMA Doubleday.
Sayigh, A. (ed.) (2019), *Sustainable Vernacular Architecture: How the Past can Enrich the Future*, Cham: Springer Nature Switzerland AG.

Schoenauer, N. (2000), *6000 Years of Housing*. Revised and expanded edition. New York: Norton.
Steyn, G. (2006), "The indigenous rondavel. A case for conservation", *South African Journal of Art History*, 21, n. 1, pp. 21–38. Available online: https://repository.up.ac.za/bitstream/handle/2263/10792/Styen_Indigenous(2006).pdf [access on 16th April 2019].
Taton, D. (1984), *Architecture traditionnelle au Niger*, U.N.E.S.C.O. Available online: https://unesdoc.unesco.org/in/documentViewer.xhtml?v=2.1.196&i...=en&multi=true&ark=/ark:/48223/pf0000025219/PDF/025219freb.pdf [access on 5th April 2019].
Tzonis, A., Lefaivre, L. (1981), "The grid in the pathway: an introduction to the work of Dimitris and Suzana Antonakakis", *Architecture in Greece*, n. 15, pp. 164–178.
Upton, D. (1983), "Power of things: recent studies in American vernacular architecture", *American Quarterly*, n. 35, pp. 262–279.
Vellinga, M. (2005), "Engaging the future: Vernacular Architecture Studies in the Twenty-First Century", in *Vernacular Architecture in the 21st Century: Theory, Education and Practice*, Asquith, L., Vellinga, M. (eds), Abingdon: Taylor & Francis, pp. 81–94.
Vellinga, M. (2014), "'Built to meet needs': Paul Oliver and the study of vernacular architecture", in *Vernacular Heritage and Earthen Architecture: Contributions for Sustainable Development*, Correia, M, Carlos, G., Rocha, S. (eds), London: Taylor & Francis Group, pp. 3–7.

Vellinga, M. (2015), "Vernacular architecture and sustainability: two or three lessons ...", in *Vernacular Architecture: Towards a Sustainable Future*, Mileto, C., Vegas, F., García Soriano, L., Cristini, V. (eds), London: Taylor & Francis Group, pp. 3–8.
Walton, J. (1956), *African Village*, Pretoria: Van Schaik.
Weber, W., Yannas, S. (2014), *Lessons from Vernacular Architecture*, Abingdon: Routledge.
Wells, C. (1986), "Old claims and new demands: vernacular architecture studies today", *Perspectives in Vernacular Architecture*, 2, pp. 1–10.
Wines, J., Jodidio, P. (2000), *Green Architecture*, Köln: Taschen.
Zakari, M. (2009), *Islam dans l'espace nigérien: 1960–2000*, Paris: Harmattan.
Zeleza, P. T. (1993), *A Modern Economic History of Africa: The Nineteenth Century*, Vol. 1, Nairobi: English Press Ltd.

>>>Ch [03]

Banham, R. (1966), *The New Brutalism*, London: Architectural Press.
Betsky, A. (1998), "Azienda vinicola Dominus, Yountville, California", in *Domus*, 803, pp. 8–17.
Brownell, B. (2014), Herzog & de Meuron Shapes A Processing Plant with Rammed Earth. *Architect Magazine*. Available online: https://www.architectmagazine.com/technology/herzog-de-meuron-shapes-a-processing-plant-with-rammed-earth_o [accessed on 1st March 2020].

Caravatti, E., Caravatti, M. (2006), "Repubblica del Mali, Africa Occidentale, scuola elementare a N'Tyeani" (trad. "Republic of Mali, West Africa, Elementary School in N'Tyeani"), in *Casabella Almanacco di Architetti Italiani*, Milan: Electa, pp. 51–57.

Cornell, D., Lima, Z. R. (2017), *Albert Frey and Lina Bo Bardi: A Search for Living Architecture*, Palm Spring: Palm Spring Art Museum and Delmonico Book.

Derrida, J. (1967), *On Grammatology*, Paris: Les Edition de Minuit.

Espuelas, F. (2009), *Madre Materia*, Milan: Marinotti.

Fathy, H. (1969), *Architecture for the Poor*, Cairo: the American University in Cairo Press.

Higgins, H. B. (2009), *The Grid Book*, Cambridge: MIT Press.

Kalili, N. (1996), *Ceramic Houses and Earth Architecture: How to Build Your Own*, Hesperia: Cal-Earth Press.

Kapfinger, O. (2010), *Haus Rauch: A Model of Advanced Clay Architecture*, Basel: Birkhauser Architecture.

Kunsmann, J. (2012), "Utopia di terra", *Domus*, 962, pp. 37–45.

Loos, A. (1908), *Ornament und verbrechen*, Vienna: Metro Verlag Edition.

Lo Ricco, G., Micheli, S. (2003), *Lo spettacolo dell'architettura. Profilo dell'archistar*, Milan: Mondadori.

Manzini, E. (1986), *La materia dell'invenzione. Materiali e progetto*, Milan: Arcadia Edition.

Oddo, F. (2015), "Materiali e imprese locali per il nuovo centro riabilitativo firmato in Mali da Emilio e Matteo Caravatti", in *Edilizia e territorio del sole 24h* (3rd December 2015). Available online: http://www.ediliziaeterritorio.ilsole24ore.com/art/progettazione-e-architettura/2015-11-23/emilio-e-matteo-caravatti-non-dimenticare-che-marginalita-e-due-passi-noi-132128.php?uuid=ACCuNdfB [access on 4th June 2019].

Pantaleo, R. (2016), *La sporca bellezza*, Milan: Elèuthera editrice.

Rocca, A. (2010), *Low Cost Low Tech, Inventions and Strategies*, Schio: Sassi editore, pp. 131–140.

Reynolds, M. (1990), *Earthship: how to build your own, vol. 1*, New Mexico:solar survival architecture.

Schumacher, E. F. (1973), *Small is beautiful: a study of economics as if people mattered*, New York: Blond & Briggs.

Sinclair, C. (2006), *Design Like You Give A Damn: Architectural Reponses to Humanitarian Crises*, New York: Thames & Hudson pp. 250–255.

Snozzi, L., Merlini, F. (2014), *Architettura inefficiente*, Novara: Edizioni Sottoscala.

Zabalbeascoa, A. (2019), "Bigger better and still affordable", *Domus*, vol. 1034, pp. 424–429.

Zevi, B. (1994), *Architettura della Modernità*, Roma: Newton Compton Press.

>>>Ch [04]

Agricola, G. (2004), *De natura fossilium (Textbook of mineralogy)*. Translated from the first Latin edition of 1546 by Mark Chance Bandy and Jean A. Bandy. Mineola, NY: Dover Publications.

Aime, M. (2000), *Diario Dogon*, Torino: Bollati Boringhieri.

Asquith, L., Vellinga, M. (2006), *Vernacular Architecture in the Twenty-first Century. Theory, Education and Practice*, Abingdon: Taylor & Francis.

Austin, G. (1984), "Adobe as a building material", *New Mexico Geology*, November, Socorro: New Mexico, pp. 69–71.

Baumann, Z. (1999), *La società dell'incertezza*, Bologna: Il Mulino.

Baumann, Z. (2001), *Voglia di comunità*, Roma-Bari: Laterza.

Baumann, Z. (2002), *Modernità liquida*, Roma-Bari: Laterza.

Capponi, M. (2008), *Un giocattolo per la mente. L'"informatica cognitiva" di Seymour Papert*, Perugia: Marlocchi.

Cointeraux, F. (1791) *L'école d'architecture rurale, ou Leçons par lesquelles on apprendra soi-même à bâtir solidement les maisons de plusieurs étages, avec la terre seule ou autres matériaux les plus communs et du plus vil prix*, Lyon: l'École d'architecture rurale.

Di Venosa, M., Morrica, M. (2018), *Rigenerare territori fragili. Strategie e progetti*, Rome: Aracne.

Davis, M., (2006), *Planet of Slum*, London: Verso.

Dieste, E. (1996), *Architettura, partecipazione sociale e tecnologie appropriate*, Milan: Jaca Book.

Fathy, H. (1969), *Architecture for the Poor*, Cairo: The American University in Cairo Press.

Frampton, K. (2001), *Studies in Tectonic Culture: The Poetics of Construction in Nineteenth and Twentieth Century Architecture*, Cambridge, MA: MIT Press.

Garg, A., Yalawar, A., Kamath, A., Vina, J. (2014), "Effect of varying cement proportions on properties of compressed stabilized earth blocks (CSEB): A sustainable low-cost housing material". Conference paper: *International Conference on Sustainable Civil Infrastructure*, ASCE India Section 17–18 October 2014, Hyderabad, India. Available online: https://www.researchgate.net/publication/267637559_Effect_Of_Varying_Cement_Proportions_On_Properties_Of_Compressed_Stabilized_Earth_Blocks_CSEB_-_A_Sustainable_Low-Cost_Housing_Material/citation/download [access on 2nd May 2019].

Goodhew, S., Griffiths, R. (2005), "Sustainable earth walls to meet the building regulations", *Energy and Buildings*. 37 n. 5, pp. 451–459. Available online: https://www.researchgate.net/publication/223843448_Sustainable_earth_walls_to_meet_the_building_regulations [access on 20th June 2019].

Houben, H., Guillard, H. (1994), *Earth Construction: A Comprehensive Guide*. London: Intermediate Technology Publications.

Houben, H., Guillaud, H. (1998), *Traité de construction en terre*, Marsiglia: Editions Parenthèses.

Kalili, N. (1996), *Ceramic Houses and Earth Architecture: How to Build Your Own*, Hesperia: Cal-Earth Press.

Keefe, L. (2005), *Earth Building: Methods and Materials, Repair and Conservation*. London: Taylor & Francis.

La Touche, S. (2007), *Petit traité de la décroissance sereine*, Paris: Mille et Une Nuits.

Levi-Strauss, C. (1964), *La Pensée sauvage*, Milan: Il Saggiatore.

Marchand, T. (2009), *The Masons of Djenne*. Bloomington: University of Indiana Press.

Martinotti, G. (1993), *Metropoli. La nuova morfologia sociale della città*, Bologna: Il Mulino.

McArdle, P. (2011), "Afghanistan's last locavores", *The New York Times*, 19th June.

Michael, A., Philokyprou, M., Thravalou, S., Ioannou, I. (2016), "The role of adobes in the thermal performance of vernacular dwellings", in *Terra Lyon 2016: Articles Selected for On-Line Publication*, Thierry, J., Hubert, G., Chamsia, S. (eds.), Villefontaine: CRAterre. Available online https://craterre.hypotheses.org/1827.

Minke, G. (2006), *Building with Earth*: Basilea: Birkhauser.

Morgan, W.N. (2008), *Earth Architecture: From Ancient to Modern*, Gainesville: University Press of Florida.

Norberg Schulz, C. (1979), *Genius Loci: Towards a Phenomenology of Architecture*, Segrate: Rizzoli.

Oliver, P. (1997), *Encyclopedia of Vernacular Architecture of the World*, Cambridge: Cambridge University Press.

Önday, Ö. (2019), "Japan's Society 5.0: going beyond Industry 4", *Business and Economy Journal*. Available online: https://www.researchgate.net/publication/335763887_Technology_trends_Education_40_and_beyond [access on 20th May 2019].

Papert, S. (1993), *The Children's Machine: Rethinking School in The Age of the Computer*, New York: Basic Books.

Petrillo, A. (2006), *Villaggi, città, megalopolis*, Rome: Carocci.

Pignatti, I., Rovigatti, P., Angelucci, F. (2018), Territori fragili. *Paesaggi città, architetture, IFAU 2018 – 2nd International Forum on Architecture and Urbanism*, Rome: Gangemi.

Pumpelly, R. (1908), *Explorations in Turkestan: with an account of the basin of Eastern Persia and Sistan. Expedition of 1903, Vol. 26*, Washington: Carnegie Institution of Washington. Available online: http://www.etana.org/sites/default/files/coretexts/20373.pdf [access on 5th May 2019].

Routledge Hill, D. (1996), "Engineering", in *Encyclopedia of the History of Arabic Science*, Vol 3, Rashed, R., Morelon, R. (eds.), London and New York: Routledge, p. 766.

Rudofsky, B. (1964), *Architecture Without Architects: A Short Introduction to Non-pedigreed Architecture*, Doubleday, Garden City, New York: The Museum of Modern Art.

Saxena, M., Vyas, S. (2019), "Periurban areas: sustainability dimension", in *International Journal of Mechanical and Production Engineering Research and Development (IJMPERD)*, Vol. 9, n. 2, pp. 545–554.

Schumacher, E. F. (1973), *Small is Beautiful: A Study of Economics as if People Mattered*, New York: Blond & Briggs.

Sennett, R. (1998), *The Corrosion of Character: The Personal Consequences of Work in The New Capitalism*, New York & London: Norton.
Sennett, R. (2008), *The Craftsman*, New Haven & London: Yale University Press.
Snell, C., Callahan, T. (2009), *Building Green: A Complete How-to Guide to Alternative Building Methods: Earth Plaster, Straw Bale, Cordwood, Cob, Living Roofs*, New York: Sterling Publishing Company.
Thierry, J., Hubert, G., Chamsia, S. (2018), *Terra Lyon 2016: Articles Selected for On-Line Publication*, Villefontaine: CRAterre. Available online: https://craterre.hypotheses.org/1827.
Turner, J. C. (1996), *Housing by People: Towards Autonomy in Building Environments*, Milan: Jaca Book.

>>>Ch [05]

Burkner, H. J. (2010), *Vulnerabilität und Resilienz. Forschungsstand und sozialwissenschaftliche Untersuchungsperspektiven*, Working Paper, Erkner, Leibniz-Institut für Regionalentwicklung und Struk-turplanung. Available online: http://www.irs-net.de/download/wp_vr.pdf [access on 30th May 2020].
D'Onofrio, V. (2015), "Il Delta del Niger, tra sfruttamento delle risorse, povertà e degrado ambientale", in *Notizie Geopolitiche. Quotidiano indipendente online di geopolitica e politica estera*, 12th December 2015. Available online: https://www.notiziegeopolitiche.net/il-delta-del-niger-tra-sfruttamento-delle-risorse-poverta-e-degrado-ambientale/ [access on 20th May 2020].
Frampton, K. (1995), *Studies in Tectonic Culture: The Poetics of Construction in Nineteenth and Twentieth Century Architecture*, Cambridge, MA: The MIT Press.
Garzanti Linguistica (2015), Resilience. Available online: https://www.garzantilinguistica.it/ricerca/?q=resilienza [accessed on 15th May 2020].
Jha, A. K., Miner, T. W., Stanton-Geddes, Z. (2013), *Building Urban Resilience: Principles, Tools and Practice*, Washington: The World Bank.
Langeveld, M. (2013), *The Resilient Design Principles*. The Resilient Design Institute. Available online: http://www.resilientdesign.org/?s=resilient+design+principles&x=0&y=0 [access on 30th May 2020].
Norberg Schulz, C. (1979), *Genius Loci. Towards a Phenomenology of Architecture*, New York: Rizzoli.
Petzold, H. G., Muller, L. (2004), "Resilienz und protektive Faktoren im Alter und ihre Bedeutung für den Social Support und die Psychotherapie bei älteren Menschen", in *Mit alten Menschen arbeiten*, Petzold, H. G. (ed,). Stuttgart: Pfeiffer bei Klett-Cotta, pp. 108–154.
The Resilience Alliance (2002), *The Characteristics of Resilience*. Available online: http://www.resalliance.org/index.php/resilience [access on 30th May 2020].

Along the Niger river, Gaya, Niger ©AT 2016

For Product Safety Concerns and Information please contact our EU
representative GPSR@taylorandfrancis.com
Taylor & Francis Verlag GmbH, Kaufingerstraße 24, 80331 München, Germany

www.ingramcontent.com/pod-product-compliance
Ingram Content Group UK Ltd.
Pitfield, Milton Keynes, MK11 3LW, UK
UKHW022154190926
475702UK00003B/29
* 9 7 8 0 3 6 7 7 4 4 0 3 8 *